D0952686

FRIENDSHIP

ALSO BY LYDIA DENWORTH

I Can Hear You Whisper:
An Intimate Journey through the Science of Sound and Language

Toxic Truth: A Scientist, a Doctor, and the Battle over Lead

FRIENDSHIP

THE EVOLUTION, BIOLOGY, AND
EXTRAORDINARY POWER OF LIFE'S
FUNDAMENTAL BOND

Lydia Denworth

W. W. NORTON & COMPANY
Independent Publishers Since 1923

For information about permission to reproduce selections from this book, write to
Permissions, W. W. Norton & Company, Inc., 500 Fifth Avenue, New York, NY 10110

For information about special discounts for bulk purchases, please contact
W. W. Norton Special Sales at specialsales@wwnorton.com or 800-233-4830

Manufacturing by Lake Book
Book design by Chris Welch
Production manager: Anna Oler

Library of Congress Cataloging-in-Publication Data

Names: Denworth, Lydia, 1966- author.
Title: Friendship : the evolution, biology, and extraordinary power of life's fundamental
bond / Lydia Denworth.
Description: First edition. | New York, NY : W.W. Norton & Company, [2020] |
Includes bibliographical references and index.
Identifiers: LCCN 2019032111 | ISBN 9780393651546 (hardcover) |
ISBN 9780393651553 (epub)
Subjects: LCSH: Friendship. | Friendship—Health aspects. | Neuropsychology. |
Mind and body.
Classification: LCC BF575.F66 D46 2020 | DDC 158.2/5—dc23
LC record available at https://lccn.loc.gov/2019032111

W. W. Norton & Company, Inc., 500 Fifth Avenue, New York, N.Y. 10110
www.wwnorton.com

W. W. Norton & Company Ltd., 15 Carlisle Street, London W1D 3BS

3 4 5 6 7 8 9 0

For Mark

Contents

INTRODUCTION: A New Science   1

ONE: Fierce Attachment   20

TWO: Building a Social Brain   43

THREE: Friendship under the Skin   65

FOUR: Middle School Is about Lunch   91

FIVE: A Deep Wish for Friendship   116

SIX: The Circles of Friendship   138

SEVEN: Digital Friendship   163

EIGHT: Born to Be Friendly?   183

NINE: Deeply Built into the Brain   206

TEN: The Good Life, Revealed   229

Acknowledgments   253

Notes   257

Index   279

FRIENDSHIP

A New Science

The island of Cayo Santiago is close enough to the east coast of Puerto Rico that a strong swimmer could reach it. Its southern end rises out of the sea, invitingly green, and there's a narrow isthmus in the middle studded with palm trees that would make an excellent landing spot for day trippers from the nearby fishing town of Punta Santiago. But most locals have never set foot on Cayo and never will.[1]

I arrived at the dock in Punta early one summer morning in 2016, the year before Hurricane Maria swept through and devastated the area. I was there to catch the seven o'clock boat to the island. The local fishermen, for whom Cayo remains a little mysterious, eyed the assembling group with interest. A dozen or so people waited for the Boston Whaler that would ferry them over. Half wore matching navy-blue shirts and pants identifying them as staff of the Caribbean Primate Research Center. The others wore a more unofficial uniform—long pants, T-shirts, and hiking boots—and carried backpacks hung with water bottles and large-brimmed hats to protect against the sun. They were scientists and graduate students dressed for work. Except for the occasional carefully vetted visitor like me, these are the only people allowed on the island.

The seven-minute ride was all outboard hum and light sea spray. Approaching Cayo from the water, it is possible to see what's not visible from Punta's beach. This is an island of monkeys. Suddenly they're there, spots of brown and gray perched in the trees or sitting in the shade or lying on the rocky cliffs. Occasionally they even play in the water, swinging off tree limbs and dropping into the waves like teenagers at a swimming hole.

I climbed off the boat and followed my host, Lauren Brent, a biologist from the University of Exeter in England. She dipped the soles of her boots in a bucket of disinfectant, then shouldered her backpack and I did, too. With a long brown ponytail tucked under her hat, Brent looked younger than her thirty-six years. But she was one of the people in charge on Cayo, overseeing many of the research assistants who clambered onto the dock behind us.

At the end of the dock, a sign commanded in English and Spanish: DO NOT ENTER. DANGER. THESE MONKEYS BITE! A musty smell of earth and wet fur intensified whenever the breeze died down. The chirping of birds rose and fell but never disappeared, and I realized it was not just birds I was hearing. The monkeys called to one another, at first like soft-spoken seagulls, then a sudden crescendo of screeches filled the air for a few minutes before abating. I saw why one of Brent's former assistants called the island, "a Jurassic Park kind of place."[2]

A certain vigilance is in order on Cayo. We humans are wildly outnumbered. Rhesus macaques aren't much bigger than house cats, and they reminded me of cats as they walked by, long legged and round bellied, tails waving in the air. But these monkeys are not going to curl up in your lap and purr. They are aggressive, as obsessed with status as Napoleon Bonaparte.

We hadn't gone more than a few yards when I, the newcomer, was tested by a male who refused to move from the path in front of me. I followed instructions and kept my head down, avoiding eye contact, until he moved on.

A group of juveniles peered down at us from the top of a nearby tree.

"*Hola*, I see you," Giselle Caraballo, the colony's manager, called to them with a smile.

She turned to me. Her businesslike demeanor couldn't hide her affection for the animals.

"The little ones are curious."

In a trick learned from their elders, the young macaques responded to her greeting by shaking the tree a little to show they weren't scared.

Brent chimed in. "Watch overhead." Streams of urine periodically cascade from the tree branches and anyone who spends more than a few weeks on the island gets peed on. Hats and sunglasses don't just protect against the sun.

The risk runs both ways. If humans carry disease ashore, it can be deadly for the monkeys. Tuberculosis is particularly dangerous and every human here has been tested and cleared. On the other hand, the macaques naturally carry herpes B. It's not a problem for them but can be fatal in the human nervous system: exposure via a bite, a scratch, or pee in the eye means immediate evacuation to an emergency room to begin a course of antiviral medication.

Cayo's regulars take all of this in stride. Caraballo has been working here in various capacities for about a dozen years, since she graduated from the University of Puerto Rico, which now manages the place through its Primate Research Center. When Brent is on the island, she carries herself with a no-fuss confidence earned through experience. She first arrived in 2006 to spend a year with the monkeys as she researched her dissertation, and has since spent hundreds of hours on Cayo, weathered storms and disease outbreaks, and argued with customs officials over the declared value of fecal samples she sent back to the lab. ("But it's literally shit!")[3] These days Brent leads an ambitious field research project on Cayo with her collaborator, and former boss, neurobiologist Michael Platt of the University of Pennsylvania.

Both women also know that although the monkeys are wary of humans, they are mainly interested in one another. That is why Brent works here among the exceedingly social macaques. She studies friendship, asking fundamental questions about how and why bonds develop between some individuals but not others, and measuring the consequences of those bonds.

What, she wants to know, are friends really for?

I am asking the same question. That's what brought me to Cayo Santiago. For several years now, I have been visiting the front lines of the science of friendship as well as revisiting its history. Cayo qualifies as both.

As a science writer, my usual beat is the brain. I've been following efforts to map the trillions of connections between brain cells.[4] But recently, it has become evident that there is an equally important map to be made extending out from the brain. Invisible but essential, it's the web of connections we forge with others, the network of individuals whose actions and emotions affect us just as we affect them. We may be separate beings, but we are deeply bound, as if there truly were silken threads tying us all together physiologically. Our personal webs of connection include our family members, our romantic partners, and our friends.

Of these three categories, the first two have been closely examined and much has been written about them. Rightly so. Family and romantic relationships hold enormous power over us and reverberate through our lives for good and for ill. When these bonds are happy, they bring us great joy and fulfillment. When they are unhappy, they make us miserable, sometimes crippling us psychologically. Always, we feel the footprints of our loved ones as they walk through our lives. I have spent most of the last decade supporting my husband in a new business venture, raising teenagers, and caring for a parent with Alzheimer's disease. I know all too well how much my family members' mood swings influence my own mood, and undoubtedly tip over into my physiology, at times causing my adrenaline to spike and my pulse to pound.

Friendship has the same power. It serves an elemental need to belong and triggers physical pleasure and pain to make us pay attention to that need. That's why I still remember the pure joy of riding in a convertible with my college roommate singing "Take on Me" at the top of our lungs. On the other hand, not getting invited to a birthday party (at age ten or age fifty) can feel like a betrayal on the level of Brutus and Julius Caesar.

Yet by comparison with relationships forged in blood and love, science has historically given friendship short shrift.[5] Archaeologists studying our earliest ancestors focused on bones and tools, not social lives. Biologists ignored friendships because unlike romantic or mating relationships they were thought not to affect reproductive success. The handful of psychologists and sociologists who addressed friendship were lonely, professionally speaking, though their work seems prescient today. Friendship was too ephemeral, a little too squishy, too difficult to define and measure to be taken seriously by the wider scientific world. The subject was left mostly to philosophers, and even many of them steered clear. For centuries it was considered purely cultural, an invention of human society—and modern human society at that. So much so that C. S. Lewis once wrote, "Friendship is unnecessary, like philosophy, like art. . . . It has no survival value, rather it is one of those things which give value to survival."[6]

Most of us are as guilty as scientists of failing to take friendship as seriously as it deserves. We pay lip service to it but prioritize family and romance, ditching our friends when we fall in love, or letting time with them be the first thing to go when we get busy. For years, whenever I bumped into friends on the streets of my Brooklyn neighborhood, I was quick to say, "Let's get together . . . really, let's." And I meant it. But work and family would swallow me up and I would rarely follow through. Then my family moved to Hong Kong for a few years where I knew no one. The first time I was invited to dinner with some other women I was as excited as if I'd been asked to the prom, only to be crushed when the conversation revolved around shopping and the help (domestic help is the rule not the exception in Asia). I went home and wept to my husband, "These are not my people." It got better in time—I did make some good new friends, and was joined in the second year by an old one—but it was obvious how much I had taken for granted my friends at home. My regrets were as searing as my loneliness.

Each of us is constrained by time. But we might want to rethink how we apportion the time we have. All due respect to C. S. Lewis, but he was wrong. It turns out that friendship does have survival value in the most literal sense—more socially integrated people live longer than those who are less well connected.

With this new understanding—built incrementally because that is how science proceeds—a different breed of scientist, the sort who thinks about health and biology and evolution, is approaching friendship with new respect. It is now recognized as a critical piece of social behavior that has been hiding in plain sight. Yes, friends are one of the great pleasures of life (Lewis was right there), and our bonds with them are undeniably shaped by culture. But that is clearly only part of the story. There are biological and evolutionary foundations to friendship.

The discovery and exploration of those foundations has clarified what friendship is. It has helped to determine where friends are distinct from other relationships, like spouses or sisters, and where those lines of demarcation blur. It has tipped the scale toward quality rather than quantity of relationships, though not entirely. It has begun to shed light on the underpinnings of the social dysfunction that accompanies neurological conditions like autism. And critically, the new science of social behavior has also starkly revealed the physiological consequences of friendship's opposite—loneliness—which is as deadly as cigarette smoking or obesity.

The busy, complicated social lives of Cayo's macaques are one of the places researchers have been digging for the roots of friendship. That may seem unlikely but in this remarkably interdisciplinary field, some of the most important breakthroughs have come in studies of animals. This is not to say that every animal is equally relevant to human beings, but neither are they irrelevant. Friendship or something approaching it has recently been identified in a surprising number of other creatures, from dolphins to zebras.[7] Even zebra fish exhibit interesting social behavior. They show reduced levels of fear when they can smell famil-iar fish and even lower levels if they can see their "friends."[8] And I was quite taken by the news that sheep are able to recognize animals they grew up with after years of separation.[9] "As we think deeply about what friendships are, we're starting to find them in other species," Brent says. "That means there's a story there that goes beyond humans and human society."

There is a lively debate about the validity of comparisons between humans and other animals. Anthropomorphism, the sin of imput-ing human intentions and ideas to nonhuman creatures, is always a concern. But so, these days, is what primatologist Frans de Waal pro-vocatively calls *anthropodenial*, which he defines as an unwillingness to believe in the extent of other species' demonstrated capabilities.[10] Among primatologists, at least, it is no longer forbidden to call an animal's friend a friend—at least in casual conversation—rather than a "preferred social partner." And Brent's generation, standing on the shoulders of those who went before, can now unabashedly admit to studying the science of friendship.

Animals are only part of the story of the biology and evolution of friendship, of course. Humans have also proven to be full of surprises. We've known for a while that the presence of a friend can lower our stress response and make it easier to tackle a challenge.[11] Now that we can better decipher our brain waves and genetic profiles, we've also learned that they reveal quite a bit about our relationships. By looking at the way two people's brains process video clips—the neu-ral version of laughing at the same jokes—neuroscientists can predict how likely those people are to be friends.[12] On the health front, our blood pressure and immune cells are strongly affected by how much we like the people with whom we spend our time. You are not imagin-ing the calming effect of a close friend. Even something as seemingly prosaic as the number of hours of togetherness required before you consider someone a close friend turns out to have a possible evolution-ary explanation.

For both humans and animals, to be social is to behave in a way that affects another being. That behavior can be positive or negative. It can be subtle and small—a glance or a touch or a whisper. It can be big and booming and brawny, like a physical attack or a loud declaration. It can be anything in between. If primates, which, of course, includes us, are specialized for anything, it's social behavior. We are gregarious. In scientific terms, that doesn't mean we all swing from the chandeliers or host never-ending parties, it simply means we are social creatures. We have always lived in groups and as those groups got bigger and more complex, we had to figure out how to navigate them and get along with one another. We had to know how to read emotions and recognize allies. As a group, we had to work out how to communicate, find food, and defend ourselves against predators. We had to learn to cooperate. As individuals, we had to know whom we could rely on in tough times and relax with in quiet times. In other words, we had to have friends.

Granted, the particulars of any animal's social behavior differ from yours and mine. Rhesus macaques can't crack jokes like my friend Moira did on the day, soon after we met, when I realized she was someone I wanted to know better. We were working as reporters in London and I was frustrated by the glacial speed of the decrepit office printer. She suggested there were elves in Victorian green eyeshades living inside it, cranking out the copy. It was a small joke, wry and unexpected, and it amused me. A few weeks later, I told Moira about a grave accident that befell my twin brother when we were children, sharing not just the facts of it but my feelings about it and the way it had changed the dynamics of my family. She felt the weight of being entrusted with my story. The best friendships invite vulnerability. Twenty-five years later, Moira and I are still helping each other laugh at adversity and telling each other our stories—the things that happened forty years ago and the things that happened four hours ago.

Animals cannot do any of that and no one is claiming they can. But animals have other tactics to deploy to build bonds—their own version of laughter and self-disclosure. Or you could argue, as a few scientists do, that laughter and sharing personal stories are our version of grooming and hanging out in close proximity.[13] Either way, this common sociability reveals not only the unexpected capacities of animals but also the unexpected roots of human affiliative behavior.

While being friendly means pulling from the kind and supportive end of one's social repertoire, it's fair to say that on Cayo, acts of good will aren't the first thing you notice about the way the monkeys interact.[14] Rather, it is the strict hierarchy that, well, dominates. That's one reason it took so long for anyone to appreciate the power of the positive interactions that occur. Macaque society is not a democracy. The island's population hovers around fifteen hundred and the animals divide themselves into distinct groups organized along matrilineal lines. Females inherit their rank from their mothers at birth and, barring rare and dramatic overthrows, stay in the same group, at roughly the same rank, for life. Daughters rank below their mothers, but above any of their older sisters so that the youngest daughter is always the most senior sibling. Males, too, get their relative rank from their mothers, but males typically outrank females. As young adults, however, males, in the biologically driven protection against incest, are expected to move on from the group in which they were born and out to new groups where they have to start from the bottom and try to work their way into power.[15]

The caste system was immediately apparent when we reached the Lower Corral, the first of three feeding areas on the island. Chain-link fence had been strung along sturdy pipe frames and strengthened with corrugated sheet metal panels to create a large contained area with a food bin inside. Hundreds of monkeys were arrayed within, atop, and outside of the fence. This was the source of the clanging and banging of metal—from the flinging of trash can lids and shaking of sheet metal—that is as much a part of the Cayo soundtrack as the calling of birds and macaques.

The caretaking staff had just filled the feeder with monkey chow, as they do first thing every morning, and two macaques, a male and a female, were standing atop the concrete base of the feeder. They had lifted the doors and were picking through the chow, tossing rejected pieces onto the ground at their feet. Occasionally, they nibbled on one that passed muster, while forty or fifty monkeys fidgeted on the ground nearby and a larger group lurked outside the corral.

"They're picky about their food," said Caraballo. "They'll look through and find that perfect piece of monkey chow. This is a luxury that higher-ranking groups get. The lower you are in the hierarchy, you have to eat whatever's left if you want to eat anything. You'll have monkeys that will eat stuff that's been half chewed up and sitting in the mud."

Clipboard in hand, Caraballo went to work, making notes about the monkeys as we talked. Each morning, she and her staff spread out through the island to survey the monkeys, taking attendance. As a result, Caraballo knows most of the monkeys well.

"He's the alpha male, 07D, and he's a bit of a jerk," she said of the bossy male at the center of the corral. "He's always chasing someone. That's the alpha female he's protecting. Basically, they have taken over the feeder. They can sit there and eat all they want. Everybody else is just sitting around waiting for them to be done."

Suddenly a scrappy youngster scurried by, loaded down with dirt-scavenged chow. I counted seven pieces, clutched in his hands and feet and stuffed into his cheek pouches. He carried his loot behind a rock to eat, though he had to scrabble with some of the monkeys on the edge of the circle to hang on to the food.

There was another feeder at the top of the hill in the Upper Corral. More than one group used this one, and the hierarchy separated them as effectively as an invisible electric fence. You could tell a monkey's power from its position, like the audience at the State of the Union address. The groups were arranged in concentric circles with the top group closest to the feeder, the next sitting just outside the fence, and the lowest-ranking animals hovering on the edges of the clearing, prepared to wait an hour or more to eat. I thought they looked resigned, sitting with their hands on their knees like people around a campfire.

Small as it is, less than forty acres in total, Cayo is as demarcated as a gang-ridden neighborhood. Stray into the wrong territory, and there could well be trouble. Group F—each is identified by letters—pretty much runs the place. They control the Lower Corral and coveted areas like the shady hillsides of Big Cay. Sometimes they wander into the mangroves that fill the low spot near the isthmus. A rocky beach forms a bridge to Small Cay, home to Group V. It's hotter and muggier over there, yet it has its own appeal. Life is calmer and less competitive and attracts the monkeys that can't make it elsewhere. "We call Group V *zafacone*, the trash can," Caraballo told me.

Brent walked through the island with a light step and an observant eye. She was informal and casual, like her twentysomething assistants with whom she chatted easily about the bars where they had hung out the night before and the cheap local beer everyone preferred. When a two-year-old female macaque stole a piece of her equipment, Brent stomped her foot and said, "Ooh, I hate you!" But her knowledge of monkey behavior runs deep. Watching the animals with her is like visiting the Louvre and discovering the unassuming tourist next to you is an art historian.

Knowing the monkeys is essential to this enterprise. Most of the scientists here spend their days like gossip columnists keeping track of who does what to whom, and in what order. That only works if they know who's who to begin with. And it works best if you can identify monkeys from a distance, from behind, or while they're on the move, just as you might find your friend in a crowd by recognizing the back of his head. Once you know the monkeys, you are more alive to the subtleties of their interactions, and better able to see their personalities.

As the day wore on, I found I could recognize the male Brent and I had been following for the last few hours. Like his peers, he had brown fur shading to orange around the haunches, a pink wrinkled face shaped like an upside-down bicycle seat, and ears like jug handles. But

unlike all the others, he also had a kinked tail and dimples. I was starting to feel some affection for him. I was also starting to spot the kind of
behavior that most interests Brent. The aggression is what you notice
first because it's noisy and physical. Like when screeching broke out
behind us. Tail up, mouth gaping, a monkey was threatening a group of
three others. They got up and moved away so the confrontation ended
almost as quickly as it began. Elsewhere, a freckle-faced male sat on
a rock. Another male, small but with piercing eyes, approached, and
Freckles was quickly up and away. Before he went, he bared his teeth at
the monkey coming his way. "That's a fear grimace," said Brent, "a way
of showing submission." Piercing Eyes, otherwise known as 4H2, took
the coveted spot on the rock.

But something quieter was going on, too. Under a tree, a female
nursed a sleepy baby. The male sitting next to her turned and started
grooming the female, picking dirt and insects out of her fur with his
long narrow fingers. On the edge of the group, a loner slept among tree
roots. It was birthing season on Cayo and many of the females were
carrying infants against their chests. If the babies were big enough,
they wandered away from their mothers to explore. But they didn't
go far and if an aggressive monkey appeared, the mothers were quick
to snatch the babies back up. An older male played with some of the
infants, roughhousing, letting them climb on him. Three females sat
lined up like train cars, each grooming the back of the one in front. I
was struck by how much they reminded me of my nine-year-old self,
sitting in a row with my friends, braiding one another's hair.

I'd been warned about that frisson of familiarity. "Spend a year
watching the monkeys and you'll never look at humans the same way
again," says Richard Rawlins, who spent a decade on Cayo as scientist
in charge. The similarities between the macaques and people can be so
striking as to be unsettling. A male trying to make a name for himself
in the feeding corral is puffed up and pugnacious like a loud local in a
bar trying to let strangers know they're not welcome. Nervous macaques
start scratching themselves in new or uncomfortable social settings, just

like people do. When two groups of monkeys get into an altercation, they face each other in lines and move backward and forward in sync, *West Side Story*-style. "We did not come up with that," said Caraballo.

Fifty years ago, when the study of animal behavior was in its infancy, "people viewed the societies of humans as very different from the societies of other animals," said Brent. "Humans were these nice rich tapestries made up of individuals. They had free will, they had variety in their social relationships." In animals, by contrast, scientists were most concerned initially with documenting broad differences between species or between males and females, juveniles and adults, and so on. The work of the intervening decades has produced a finer-grained appreciation of just how richly diverse individual animals can be and how much those differences can affect the course of life.

After a day on Cayo, I began to pick out personalities. The monkeys could be outgoing or timid, mellow or neurotic, scrappy or laid back. Some were pretty much jerks like the alpha male in Lower Corral; others won over everyone they met. A male named Chester, who died a few years ago, lives on in memory as the most charming of all. "This guy, he was good to the females," said Caraballo. "He was very gentle with them. The other males would be chasing the females and pulling their hair and biting them. He was just all about grooming them." A personality like Chester's also amounts to a strategy, explained Brent. Be good to the females and they'll be good to you. Strong female allies help a male move up in rank in a group. Other males, like the one lording it over everyone in the feeding corral, take a more aggressive approach, fighting their way up the hierarchy.

"This is an amped-up social system," said Brent of life on Cayo Santiago. Some of the pressures of life in the wild have been removed. Food is provided so the monkeys don't have to forage (there wouldn't be enough food if they did). There are no predators to protect against. That leaves the macaques pretty much the whole day to socialize in one way or another and it makes the island a perfect laboratory for Brent and her colleagues.

It also makes it an appropriate place for me to begin to explore what the new science of friendship can tell us about ourselves.

Brent and I found our own patch of shade to watch Group S while away the afternoon. It was a quiet time of day on Cayo, and the monkeys had sorted themselves into cliques, hanging out in close proximity to one another—grooming, snacking. Grouped in twos and threes or occasionally alone, they looked like people spread out at a city park on the first sunny day of spring. A lot of them were napping, lying down under a tree or against a log. They stretched out luxuriously and unreservedly, soft white bellies up, arms splayed to the sides, heads lolling. (I'll think of this moment a few years later when I read this line from a memoir of friendship: "I'd never seen someone show that much of her soft belly and receive that much respect."[16])

The kinky-tailed monkey was sitting quietly under a bay rum tree. While we waited for his next move, our conversation roamed away from the monkeys to another kind of social scene, the one that got Brent interested in friendship in the first place.

"At university, you're in that pressure cooker," she said. She grew up in Canada and it had been some fifteen years since she graduated from Montreal's McGill University, but she has never forgotten the intensity of socializing in college. "Meeting people, making friends, developing new relationships. You're forced to evaluate who you are and the type of people you want to associate with."

It's no wonder that many studies of human friendship center on college students—in addition to being easily accessible to university-based researchers, the students are deeply engaged in building a social life. In 1961, Theodore Newcomb, one of the first social scientists to ponder what drew people together, had the clever idea of arranging a year of free housing for seventeen transfer students to the University of Michigan in exchange for permission to record their interactions and ask them how much they liked and disliked one another. Tracking the men from the first day of school until the last, and noting who became friends and who did not, Newcomb demonstrated the importance of proximity, similarity, and reciprocity in creating friendships. His study was an academic version of reality television fifty years ahead of its time.[17] It was an academic version of my own college experience, for that matter. I was randomly assigned to live in a large suite with seven other women in my first year. I've lost touch with most of them completely, but one went on to be the maid of honor at my wedding. I wonder if Newcomb would have seen that coming?

It was walking into parties and other collegiate gatherings that got Brent thinking about evolution. "I was doing a biology degree and I started watching people around me the way we would any other animal," she said. "Your social life takes up so much of your mental and physical energy. We spend so much time forming relationships or ending them. Why does it seem like life or death sometimes?" At the time, primatologists following baboons in Africa were just beginning to show that friends are, in fact, vital to well-being. Brent was hooked by the idea that who you hang out with at a party could affect not only whether you had fun for one night but how the rest of your life would unfold.

The aim of much of the research on Cayo is the same as mine: To deepen our understanding of what it is about friendship that affects the course of a life. To ask not just who hangs out with whom, but what does it get them? Biologists have two ways of framing that question. Asking how friendship works addresses the here-and-now process of socializing—the physiological attributes that drive our ability to initiate interactions and to respond to them. How do you know at a glance, for instance, if a friend is elated or exhausted? You know this because specific areas of our brains are trained from birth to pay attention to faces over other body parts or objects, and to interpret every squint and smirk. Macaques do something similar with other monkeys' yawns and grimaces, each of which telegraphs meaning, albeit not the same meaning as in humans. Then there's the warm feeling we get from a hug or an affectionate pat on the shoulder, which comes from a surge of hormones into the reward centers of the brain. That warm feeling makes us come back for more and promotes bonding. Grooming works the same magic in monkeys.

To ask why friendship works rather than how is to contemplate something more ancient and fundamental. It means exploring the reasons we, and other animals, developed such skills and tendencies in the first place. Why pay so much attention to faces? Because apparently it benefits us to do so. Focusing on faces is one of many skills that make us more socially adroit. Being socially adroit—good at creating and maintaining bonds with others—makes us more likely to survive and reproduce successfully. And therefore, according to the principles of natural selection, these skills are more likely to be handed down from one generation to the next.

Biologists speak of these two explanations—how and why—as proximate and ultimate causes for behavior. They're complementary, not competing. One is near, and one is far. One is visible, the other invisible, and must be guessed at since evolutionary theory must be studied indirectly. Evolution works over thousands of years and cannot be proved by a direct experiment in the way that theories in physics can be proved.

This book considers the visible and the invisible aspects of friendship. A friendship is an organism that shifts its shape across our life spans according to our abilities and our availability—in other words, according to how much we open ourselves to its possibilities. While there is natural variation in our taste and need for companionship, there are some universals in what draws us together or throws us apart. And there is a bottom line—a biological need for connection that must be met to achieve basic health and well-being. That is why social connection is rapidly becoming an issue of public health.

This was not always obvious, though it may seem so now. It required studies following the lives of tens of thousands of people to reveal not just that social bonds and longevity go together but that loneliness is deadly. An even grander perspective—on the order of thousands of years of evolutionary history—is the only way to see that cooperation has been as essential as competition in creating who we are today.

On the other hand, it has also been necessary to zoom in. A tighter focus allows us to make out the bumps and folds of the brain, the seat of our social lives where the intricate and complex computations that go into befriending another person take place. Examining the brain highlights the sensory nature of friendship—the central role of our vision, hearing, and touch in processing social information from the outside world and passing it on to higher-order brain areas. It reveals the changing nature of the brain's social networks from infancy through adolescence to adulthood. Getting up close is also the only way to appreciate the role of genes. They are the agents of natural selection, and—we now know—change their activity in response to the environment, including social context. And finally, a microscopic approach is the only way to see what social behavior is doing at the molecular and physiological level, how it is getting under the skin.

⸺

When I got home to Brooklyn from Puerto Rico, I found my then seventeen-year-old son, Jacob, playing NBA2K with his best friend, Christian. They were sprawled next to each other on the couch, long legs spread in front of them, controllers in hand, looking at the screen. Empty sandwich wrappers and juice bottles were strewn about, evidence of an earlier run to the deli.

There was so much that was familiar about this sight. It felt like they might not have moved in the week I was gone. They had just graduated from high school and this—punctuated by a few rounds of two-on-two basketball with my younger sons in the yard—was how they were spending their days. Upon waking up each morning, they wouldn't even wait to get out of bed to communicate. "You up?" one would text the other from under the covers. Arrangements would be made to meet on the corner and get the aforementioned breakfast sandwiches. Then they'd hit the couch.

It had been driving me nuts. On some levels, I knew that wasn't fair. The summer after senior year should be a time of rest and relaxation, a rare break between the hard work of high school and starting college. Furthermore, this was temporary. In a few weeks, Jake would be heading to our family farm in upstate New York to work for the rest of the summer. He would be mucking out the pig barn, not sitting on the couch.

Still, I couldn't stand it. The responsible adult in me rebelled. Surely their days should consist of something more than this. How was it possible to spend so many hours sitting around, doing essentially nothing?

And then it hit me. Maybe they weren't doing nothing. My time watching the macaques changed how I saw these teenage boys' behavior. Look at the proximity! They sat physically very near each other, just as bonded monkeys do. Listen to the banter! Even while playing video games, they were communicating, laughing, commenting on both the game and on life, trash-talking a little now and then. They were doing the human equivalent of grooming. More significantly, like young male macaques, they were on the verge of leaving the group they'd grown up in to make their way in the wider world. They were leaving their families, and they were leaving each other. And that wasn't going to be easy, which made this time together all the more precious.

Jake and Christian met as one-year-olds and had declared themselves best friends by the age of four. They shared classrooms, teams, and even some Halloween costumes—one year it was the Hulk, another, morph suits. Their friendship survived a two-year separation when we moved to Hong Kong and, later, Jake was there for his friend when Christian's parents divorced. Their bond struck those who knew them—parents, teachers, even their slightly jealous siblings ("I don't have a Christian," said one of my younger sons)—as the best kind of friendship. Affectionate, fun, reliable, their bond was a haven, but also a foundation from which to build up the rest of their social lives. There were no signs of jealousy. They had other friends, some mutual, others less so. But they always had the comfortable home of their friendship to come back to.

I ask them eventually what is so special about their relationship.

"There's no bullshit," Jake says. "We never argue."

Christian agrees. "Never."

They glory in their enjoyment of each other's company. "I can't think of anyone I like more," Jake says.

How did they get there? How do any of us—if we're lucky enough? Friendship is so familiar and such a part of our day-to-day lives that it's easy to assume we know all about it. We don't. And we think we appreciate it. We don't—not fully. Even I, who was spending my days reporting on friendship, talking to scientists about social interaction, still missed it. I saw the video game the boys were playing and not the visceral connection between them. It is a mistake that too many of us make too often in a variety of contexts—as parents, as spouses, as friends.

It's time to bring friendship to the foreground. To see it for what it is. It actually is a matter of life and death. It is carried in our DNA, in how we're wired. Social bonds have the power to shape the trajectories of our lives. And that means friendship is not a choice or a luxury; it's a necessity that is critical to our ability to succeed and thrive. It can even be a model against which other relationships should be measured.

Our social lives have a backstory, and it's time that story was told.

ONE

Fierce Attachment

I f you wanted to pinpoint the moment when the seeds for a new way of thinking about friendship were sown, a good candidate would be the day in February 1954 when psychiatrist John Bowlby met ethologist Robert Hinde. How appropriate that the modern science of friendship has its roots in a friendship.

Both men had been invited to speak at an academic meeting in London put on by the Royal Medico-Psychological Association (today's Royal College of Psychiatrists). They later maintained they were the "second-string," asked to step in when two of the scientific superstars of the day, Konrad Lorenz and Nikolaas Tinbergen, proved unavailable.[1] The meeting's subject was the psychobiological development of the child. That wouldn't appear to have much to do with friendship but looks can be deceiving. To appreciate friendship, you must first appreciate relationships. In their separate fields, child psychiatry and animal behavior, Bowlby, who would become famous for attachment theory, and Hinde, who became a mentor to Jane Goodall and wrote some of the first papers on social bonds in animals, were pioneers in thinking about the significance of relationships.

It wasn't, however, a foregone conclusion that they would become close. Friendship usually begins with similarity. While they were both scientifically minded Brits, Bowlby was sixteen years older when the two men met, closing in on fifty and "an Englishman of an earlier, more formal generation," in the words of Hinde, who was just thirty.[2] Bowlby was a product of the British upper classes and spent most of his early life with a nanny, then was shipped off to boarding school at a tender age. (He later told his wife he wouldn't send a dog away to school.) Hinde was comparatively middle class in a time and place where that mattered, the outdoorsy son of a family that spent their holidays hiking together. It was his childhood love of bird-watching that led to his academic career.

Bowlby spent the war years working at a center for troubled youth in London while Hinde served in the Royal Air Force. Bowlby's early work had focused on the significance of disrupted childhoods and he had found that early separation and the absence of an emotional relationship with a caregiver (usually a mother) correlated with delinquency and affectionless behavior later on. One of his first papers was memorably called "Forty-Four Juvenile Thieves: Their Characters and Home-Life."[3] Hinde watched not wild teenagers but wildlife. He had devoted fifteen hundred hours to observing chickadees in the woods outside of Oxford for his dissertation. At Cambridge, where he operated a zoology field station, he was engaged in comparative studies of different species of birds, focusing on aggressive behavior and the evolutionary importance of displays. One of his early papers was on the mobbing reaction of chaffinches to owls—it wowed Lorenz, the Austrian father of the field of ethology, the new and growing science of animal behavior.

In 1954, at the time of the RMPA meeting, Bowlby was casting around outside of psychiatry for other ways to build evidence to support his nascent ideas of the critical importance of the emotional lives of children and their relationships with their caregivers. He was excited about ethology and was searching for someone who could help him apply its ideas to his budding theory about the importance of attachment. The year before, Lorenz had given Bowlby glowing reports of

Hinde's abilities. After they gave their talks at the meeting, Bowlby invited the younger man to lunch.

What they lacked in similarity, Bowlby and Hinde made up for in shared vision. Both were curious, sharp-witted, and eager to learn. By the time they had finished the meal, Bowlby was "vastly impressed" with Hinde. He had found his guide to evolutionary thinking. Together these two ushered in a new way of thinking about the importance and evolutionary adaptiveness of connections between individuals. Attachment, they would recognize, did not just matter for babies. It also helped to explain the essential nature of friendship for people of all ages.

To see the biology underpinning friendship, you must see all that friendship is and requires. Biology is much more obvious in other kinds of social interaction, such as between a mother and her baby—pregnancy, childbirth, and nursing are as embedded in bodily processes as it gets. So is sex. No one would doubt that hormones and neurochemical signaling are part of those experiences.

Friendship is subtler. It encompasses subjects that feel intangible, such as emotion, conversation, and the inner workings of the mind. It has obvious cultural overtones that can obscure the fundamentals. Dinner parties, for example, come in many flavors—from the bacchanalian feasts of the ancients to potluck church suppers and girls'-nights-out. All are ways of socializing, but what lies behind the urge to bring people together in those ways? A more basic drive to connect.

Friendship doesn't have hard edges, allowing for wide differences in approach. Some people guard the word jealously, bestowing the title "friend" on a select few. Others use it more generously, even as a generalized term of address. "Friends, Romans, countrymen, lend me your ears," says Antony in Shakespeare's *Julius Caesar*. As a result, the lines between friends and others are often indistinct. If friends are restricted to people you're not related to and don't have sex with, you couldn't call your spouse or sister a "best friend." Yet many of us do. And with the advent of social media, the word itself—"friend"—is suddenly devalued currency, so broadly applied as to be almost meaningless. I'm "friends" on Facebook with people I knew in high school and college but might not recognize if we passed each other on the sidewalk. I'm friends with people I see almost every day and people I haven't seen in thirty years, though demonstrably not in the same way.

Just because something is hard to pin down doesn't mean it's impossible to understand. Yet scientists tend not to study what they can't easily define and measure. Many of history's great thinkers also relegated friendship to the sidelines. When Michael Pakaluk, a professor of ethics at Catholic University in Washington, D.C., set out in 1991 to gather "the central philosophical writings on friendship"—everything from Socrates through Montaigne and Ralph Waldo Emerson to a 1970 essay by the writer Elizabeth Telfer—he found that the result made for "a slim anthology."[4] Some philosophers deemed friendship uninteresting or of marginal importance. Its inherent partiality smacked of patronage or put it at odds with moral philosophy.[5] Since, by even a limited definition, I prefer a friend to someone who is not a friend, I am much more likely to do a favor for my friend than for an acquaintance. Such differentiated relationships are what friendship is all about. Nowhere is this basic principle of friendship put more baldly than in the saying: "A friend helps you move house. A good friend helps you move a body."[6]

The philosophers of Athens spent an awful lot of time hanging out with their friends, so it is fitting that they wrote compellingly about those relationships. Plato's dialogue on friendship, *Lysis*, opens with Socrates ambling along between the Academy and the Lyceum and being invited to join a group of young men standing together outside a wrestling school where they liked to congregate. "A lot of us spend our time here . . . discussing things, and we'd be glad to have you join in," the youths say to Socrates. He agrees mainly, it seems, for the fun of it. "Philosophy, as [Socrates] practiced it, is often the expression of friendship of a certain sort," Pakaluk writes.[7] On the other hand, while "passionate" about his friends, Socrates leaves the reader at a loss as to what a friend is.

Aristotle had a clearer vision of friendship, one that comes closest to what we know to be true today. His thinking on the subject was so profound that it has colored nearly all that came after. Loosely translated, the Ancient Greek word for friendship is *philia* and Aristotle thought it one of life's unalloyed pleasures. He believed that *philia* came in different flavors: for profit (a business relationship), for pleasure (romantic love), and for virtue (a true meeting of the minds). He also thought friendship, in an all-encompassing sense, necessary for all people—rich, poor, young, old, male, female, even animals (a view that wasn't revisited for thousands of years). He recognized that friendship requires positive feeling, reciprocity, and time and familiarity. "As the proverb goes, [people] cannot know each other until they have shared the traditional [peck of] salt," he wrote, "and they cannot accept each other or be friends until each appears lovable to the other and gains the other's confidence. . . . For though the wish for friendship comes quickly, friendship does not."[8] But his most provocative idea was that "a friend is another self." Neuroscience and genetics will show thousands of years later that there was more to this idea than Aristotle could possibly have known.

In the eighteenth century, the Scottish philosopher Adam Smith, author of *The Wealth of Nations* and the founder of modern economics, had an insight that would prove remarkably astute later on. He was one of the first to recognize empathy, which he called "fellow feeling" in his 1759 book *The Theory of Moral Sentiments*. He saw the emotion, in which an individual physically feels what another is feeling, as a basis for morality, and he optimistically imagined a politics based on private friendship that could bring individuals together to create a morally and economically justified society.[9] That didn't happen, to put it mildly.

When psychology was born at the end of the nineteenth century, its great pioneer William James guessed that there must be biological underpinnings to psychological processes. And he astutely recognized that social connection was not always nurtured as it should be: "Human beings are born into this little span of life of which the best thing is its friendship and intimacies, and soon their places will know them no more, and yet they leave their friendships and intimacies with no cultivation, to grow as they will by the roadside, expecting them to 'keep' by force of inertia."[10]

In the same era, the precursor to much modern understanding of the significance of social context was the work of French sociologist Émile Durkheim. He recognized that people were embedded in social groups and that those groups could materially influence well-being. He also highlighted the perils of social disconnection, which he called "anomie." In his groundbreaking 1897 book on suicide, he showed that more socially connected people were less likely to attempt suicide and pointed out that individual agency was not the only factor at work.[11] That made him one of the first people to connect the bonds of friendship explicitly to mental health.

The dominant schools of psychological thought, psychoanalysis and behaviorism, continued to ignore friendship, however. Sigmund Freud saw every relationship primarily through the filter of sex, while the behaviorists went to extremes in focusing only on what they could see and test. Both traditions had the effect of limiting appreciation of plain old affection.

⌒

Enter Bowlby. He was convinced that early relationships were of fundamental importance, but when he reviewed the existing research, he could find precious little that framed the issue the way he saw it. The dogma of the early twentieth century can be summed up by the work of John B. Watson, president of the American Psychological Association and author of the 1928 best-seller *Psychological Care of Infant and Child*. Watson wrote an entire chapter on "The Dangers of Too Much Mother Love." As science writer Deborah Blum writes, his theory amounted to this: "Too much hugging and coddling could make infancy unhappy, adolescence a nightmare—even warp the child so much that he might grow up unfit for marriage." And furthermore, Watson thought irreparable damage could be done in a matter of days.[12]

What work there was on early childhood dealt with neglect. Those studies were limited, but they were powerful. One of the first appeared in the early 1940s. A psychoanalyst named René Spitz, working with psychologist Katherine M. Wolf, decided that the best way to measure the critical importance of a mother was to watch what happens to a baby who doesn't have one. He also knew that the fate of young children living in orphanages was appalling. In some institutions, mortality rates were over 70 percent. "[Institutionalized children] practically without exception developed subsequent psychiatric disturbances and became asocial, delinquent, feeble-minded, psychotic, or problem children," Spitz wrote. A lack of stimulation and a lack of mothers seemed to be the culprits. In the name of hygiene, institutions went so far in sterilizing the children's surroundings that they "sterilized the child's psyche," he said.[13]

To study the problem, Spitz followed 164 children over the first year of life. Sixty-one of the children were cared for by nurses in a traditional orphanage. The rest lived with their parents in a variety of settings, including a prison for incarcerated mothers. Whether at home or in the prison nursery, the babies who lived with their parents all began and ended the study in roughly the same place developmentally according to the tests available at the time. The prison babies actually flourished, which Spitz attributed to the devotion of their young mothers who had little else to do but dote on their offspring. The orphanage babies, however, deteriorated dramatically. Extremely susceptible to infection and illness, they were "dying from sadness," Spitz wrote. He diagnosed them with "hospitalism," which he defined as "a vitiated condition of the body." Cut off from visual stimulation by sheets hung over their crib railings, they had no toys and lacked all human contact for most of the day. "Each baby lies in solitary

confinement up to the time when he is able to stand up in his bed."
Some never achieved that milestone. Spitz concluded that it wasn't
the lack of perceptual stimulation that accounted for the children's
deprivation. "We believe they suffer because their perceptual world
is emptied of human partners." Spitz followed the institutionalized
children for two years. Their mental development was severely delayed
and 37 percent of the original group died.[14]

When academic papers didn't spur people to action, Spitz and others
turned to the power of images. In 1947, Spitz made a grainy black-and-
white film called *Grief: A Peril in Infancy*. It introduced one baby after
another as they arrive at an orphanage happy and chubby, then over
the ensuing weeks visibly deteriorate. Unsmiling, sobbing, clinging to
Spitz, the babies appear to search for their mothers. The title card over
one poor infant read: "The cure: Give Mother Back to Baby."[15] A second film was produced in the 1950s. Made on the cheap by Scottish
medical researcher James Robertson, it was called *A Two-Year-Old Goes
to Hospital*. At the time, parents of hospitalized children were granted
only brief weekly visits. Doctors and hospital staff thought such isolation was necessary for hygienic reasons and that children didn't suffer
from it. But Robertson's film said otherwise. It followed a lovely toddler named Laura who handled her initial arrival at the hospital with
aplomb but begged her parents to take her home after one week and
then barely spoke to them the following week. "At the end of the film,
she was like a frozen child, silent and unresponsive," writes Blum.[16]

Psychiatrists trashed such efforts as overly sentimental and unscien-
tific. But Bowlby didn't. He was strongly affected by Robertson's film.
It affirmed his view that babies and toddlers needed love. In 1951 he
had delivered a blistering report to the World Health Organization.
"The mothering of a child is not something which can be arranged by
roster; it is a live human relationship which alters the characters of both
partners," he wrote. His powerful account of the link between maternal
care and mental health began to make waves, but he needed more solid
evidence and he thought he could find it in ethology.[17]

Ethology was then brand-new. Lorenz, together with Niko Tinbergen, a Dutchman who taught for years at Oxford, and Karl von Frisch, another Austrian, ushered in a radical departure in methods and goals that ultimately earned the three of them a shared Nobel Prize in 1973.[18] First, they strove to study animals in their natural habitats rather than in captivity. And second, they argued forcefully that complex patterns of behavior were as much the product of evolution as morphological structures like antlers and beaks. Von Frisch, for example, deciphered a particular flight pattern, the waggle dance of the honeybee.[19] After discovering a good food source, a bee returned to the hive and conveyed the information to its friends by flying in a figure eight oriented in the direction of the food and waggling its body to a greater or lesser extent depending on how spectacular the find was. In 1951, Tinbergen published an influential book called *The Study of Instinct* that brought the ideas of ethology to a wider audience. Bowlby, for one, pored over it. Twelve years later, Tinbergen would write an equally important essay laying out four questions that have guided the field ever since. Those four questions inform the two-layered approach of asking both how and why animals behave as they do—two questions for how and two for why—and are the basis of today's new investigations of friendship. What causes behavior (meaning what is the underlying physiology)? How does it develop over an animal's lifetime? What is its adaptive value? And how has such a behavior evolved?[20]

One of the most important early ideas in ethology was Lorenz's discovery of imprinting. This was the pattern by which baby birds learn to recognize their parents. After a series of experiments, Lorenz concluded that young birds become socially attached to the first visible "conspicuous object"—bigger than a matchbox and capable of movement (or rather, movable). Even a red box set spinning or a green ball would work as effectively as a bird's natural mother, Lorenz argued. Famously, he got ducks and geese to adopt him as their parent, which led to the iconic photographs of Lorenz being followed across a lawn by a line of young graylag geese.[21]

Bowlby immediately saw the relevance of this new way of thinking about behavior. He believed that babies behaved as they did for a reason located deep in human history. In their many "long, long discussions," Hinde recalled that he might mention to Bowlby something like the fact that "baby ducks must stay near their mother otherwise a peregrine falcon or something might get them and he picked that up and wove it into the understanding of child behavior." In Hinde, Bowlby found an editor as well as an intellectual partner. The ethologist read all of the psychiatrist's early papers on the development of attachment theory and routinely sent the drafts back covered in red comments just as he did with his many students. (In return, Bowlby dedicated one of his books to Hinde.)

Bowlby also enlisted Hinde as a collaborator in his efforts to show that those isolated children in hospitals, featured to such poignant effect in Robertson's film, were suffering real and lasting harm. At Madingley Field Station, which was up until then focused only on birds, Hinde created a captive colony of rhesus macaques expressly to study the costs of separation of infants and mothers. He found that separation did indeed bring hardship to the monkeys, but he learned that not every relationship was equivalent. The effects of separation varied according to the relationship of the mother and infant and, furthermore, those relationships varied according to the social context. Hinde began to realize that it would be critical to define what a relationship actually was and how it might be shaped by circumstances.[22] After much thought and experiment, he arrived at a definition that considered a relationship—including those between friends—the result of repeated interactions between two individuals, each interaction building upon the last. His definition captures the way a satisfying conversation makes you want to talk to someone again, and how the next interaction is richer for what came before.

The work with macaques brought Hinde into the early world of primatology where he would become an important mentor, willing to take on such unlikely students as Jane Goodall (who hadn't been to college when Louis Leakey arranged for her to get a PhD at Cambridge) and Dian Fossey, who worked with gorillas in Rwanda in unconventional ways.[23] Hinde credited both women with convincing him of the importance of individual relationships and individual differences in animals.

Hinde also introduced Bowlby to the work of a young American psychologist at the University of Wisconsin named Harry Harlow.[24] Just as Bowlby and Hinde would go on to do, Harlow thought that he could uncover some of the fundamental truths about love and affection, not by looking at human mothers and babies but by looking at monkeys. Brilliant and iconoclastic, Harlow conducted now famous experiments with rhesus macaques that were controversial even in the 1950s, but made a forceful argument. In *Love at Goon Park*, her biography of Harlow, Deborah Blum writes that the power of his work connecting love and a realized life "forced you to confront how much relationships matter in life." Harlow certainly believed that was his overarching result: "If monkeys have taught us anything," he said, "it's that you've got to learn how to love before you learn how to live."[25]

Harlow did with monkeys what you couldn't do with humans (and cannot do today with monkeys). He manipulated their upbringings, isolating them, giving them different kinds of caretakers and different kinds of playmates at different points in their lives. In his most iconic experiment, Harlow separated young monkeys from their mothers and raised them in a cage with two possible surrogates. Both "mothers" were made of wire mesh and topped with something approximating a face. The major difference was that one had a feeding milk bottle attached to its midsection; the other was covered in soft terrycloth but had no food to offer. Without fail, the young monkeys visited the wire mother only to eat. The furry mother was Mom. That meant food alone, contrary to the reigning theories of the Freudians, was not enough to create a maternal bond.

In the late 1950s, Bowlby published a paper called "The Nature of the Child's Tie to His Mother."[26] In it, he laid out the beginnings of attachment theory with its evolutionary overlay. Being loved matters.

That is the idea at the core of the theory. While babies do need other things—food, shelter, and to be kept clean and safe—what matters most is love. A newborn baby can't do much, but everything in its limited repertoire—"sucking, clinging, following, crying, and smiling"—is designed to secure love by keeping a parent close, Bowlby wrote.[27] He was explicit about the adaptive value of such behavior. Without a parent to care for it, a baby would die. Simple as that.

It's hard to overstate how revolutionary such an idea was. For a time, Bowlby became persona non grata in British psychiatric circles and stopped going to meetings. Today, he is justifiably remembered for fundamentally changing the world's understanding of the early years of life. What's less well recognized, but significant for the history of friendship, is that although Bowlby's work centered on mothers and children, he ultimately conceived of attachment as a lifelong phenomenon. In the first book in his seminal trilogy, *Attachment*, he wrote that in adolescence and adult life, attachment behavior "is a straightforward continuation" of the same behavior in childhood and was commonly directed outside the family, even toward groups and institutions. "In sickness and calamity, adults often become demanding of others; in conditions of sudden danger or disaster a person will almost certainly seek proximity to another known and trusted person." He was offended by the Freudian view that such behavior was unnatural. "To dub attachment behavior in adult life regressive is indeed to overlook the vital role that it plays in the life of man from the cradle to the grave."[28]

A few short years after Bowlby and Hinde met, another important friendship in the science of friendship blossomed, this one in Massachusetts. In 1956 a twenty-six-year-old man named Stuart Altmann knocked on the office door of biologist Edward O. Wilson at Harvard University. Altmann had a problem. A newly arrived doctoral student, tall and bearded and serious, he was something of an anomaly in the halls of Harvard because of his field of interest. After getting undergraduate and master's degrees in biology from the University of California,

Los Angeles, Altmann had spent the previous two years working on parasites for the Army Medical Service where a colleague had told him about an unusual island off the coast of Puerto Rico and the monkeys that lived there freely. Altmann was curious about social behavior. Cayo sounded to him like an excellent place for a major study of social communication that could form the basis of a dissertation. But this was a new kind of field research, rarely attempted.[29] There was virtually no one studying primate behavior outside of captivity. (The Europeans were still mainly working with birds and insects.) It would be four more years before Jane Goodall set foot in the Gombe Reserve where she would transform the way scientists, and the public, thought of chimpanzees. As a result, when Altmann went looking for an advisor, no one in Harvard's biology department was particularly interested in his ideas or knew how to provide guidance. Finally, the department chair sent him to Ed Wilson, then a junior fellow. "We're 95 percent sure he will be a faculty member next year," he told Altmann. "Go talk to him."[30]

Wilson was not even a full year older than Altmann. A passionate naturalist, he had grown up exploring the outdoor world in his native South, inspired by the first jellyfish he ever saw on a Gulf Coast beach the summer he was seven. That same summer determined the course of his life in another way. One day, while catching pinfish off a dock, Wilson yanked a bit too hard to get one off his hook. The fish flew into his right eye and the damage from its spine caused him to lose most of the vision in that eye within months. Then, as a teenager, Wilson lost most of his hearing in the upper frequencies, which he later attributed to a genetic condition. When he donned binoculars and went out birdwatching, he discovered he could neither hear nor see the birds. And so, he turned to studying insects, ants in particular. "I had to have one kind of animal if not another, because the fire had been lit and I took what I could get," he wrote in his memoir *Naturalist*. "I would thereafter celebrate the little things of the world, the animals that can be picked up between thumb and forefinger and brought close for inspection."[31]

Even as Wilson focused on the ground, on recognizing the minute details that differentiated one species of ant from another, and cataloguing the world's population of the *Formicidae* family, he was not just an entomologist. He thought of himself as a storyteller and the tale he was recounting was a grand one, using the patterns that he found in nature to piece together the history of life on Earth. "Where the experimental biologist predicts the outcome of experiments, the evolutionary biologist retrodicts the experiment already performed by Nature; he teases science out of history," Wilson wrote.[32] In his view, those who know natural history already have an abundance of answers in the facts and data they can see in front of them. "What they most need are the right questions."

The fundamental question at the heart of evolutionary science is "Why? Why are people, monkeys, birds, ants, and every other creature the way we find them?" Charles Darwin went a long way in answering that question with his theory of natural selection. He worked out that living organisms display a variety of characteristics, or traits, such as the color of a hummingbird's feathers, the size of the blue whale, or the range of hair colors evident in human beings. And that those traits that prove most useful, that provide an edge to the creatures that possess them, tend to appear in greater numbers in the subsequent generation. But there were holes in this idea. The problem that no one could solve was how exactly natural selection, the core of the Darwinian theory, would work in practice. "Heredity was one of Darwin's great frustrations," wrote Carl Zimmer in *Evolution: The Triumph of an Idea*.[33] Coincidentally, at the same time that Darwin was writing, the monk Gregor Mendel was growing peas in his garden in what is now the Czech Republic. By breeding thousands of peas, generation after generation, Mendel worked out some of the underlying principles of heredity, but no one connected his ideas to Darwin's until decades later. Without that essential link, the principle of natural selection fell so out of favor that by 1920 some no longer believed it could be true.

Then came a breakthrough. A group of scientists led by the American biologist Sewall Wright and British geneticist and statistician Ronald

Fisher finally joined natural selection with genetics. They showed that selection "was largely a matter of the changing fortunes of different forms of genes," wrote Zimmer.[34] Their work gave Darwin's theory the firm foundation it had been missing. Others built on that foundation; zoologists catalogued species and paleontologists accumulated enough of a fossil record to find meaningful patterns. By the 1950s, this deeper understanding of evolution and genetics, known as the modern synthesis, had taken hold.[35]

But there remained one more problem with Darwin's theory. What could explain altruism? Darwin had recognized that some ants in a colony fail to reproduce, devoting their lives to serving others in the group. And he wondered about birds that give warning calls when predators appear. Why make yourself an easier target? Such self-sacrifice didn't seem to fit with a theory in which the drive to survive was paramount.

This was still an open question when Altmann told Wilson that he wanted to spend two years watching and recording the social behavior of the rhesus macaques on Cayo Santiago. Wilson was enthusiastic. "We hit it off immediately," Altmann said years later. "It didn't take us very long to come to the realization that the two of us were respectively studying ants and primates, two groups of organisms that have the most complex social systems in the animal kingdom. The question was, are there some commonalities?"[36]

In June of 1956, Altmann landed for the first time on Cayo Santiago at the same spot where I arrived with Brent. The island itself was an unlikely experiment in social behavior that had begun in 1938. With World War II looming, travel to some parts of the globe was becoming dangerous or downright impossible. Clarence Carpenter was one of a handful of American scientists worried that the coming hostilities would cut off access to Old World monkeys from India and the Far East.[37] A pioneer in the study of primates in the wild, Carpenter was interested in behavioral questions and he couldn't answer those without animals to observe. His academic base, Columbia University,

was also home to the School of Tropical Medicine, whose researchers wanted a steady supply of animals for their biomedical studies.
Carpenter convinced his colleagues that the solution was to create a
new breeding colony in North America. They would stock it with rhesus macaques, a hardy breed that was useful in the laboratory. Cayo
Santiago, a small and, at the time, mostly barren island off the east
coast of Puerto Rico near the city of Humaçao, emerged as a possible
site. The island's owners, who used it for grazing goats, were persuaded
to lease it for science.

Money was raised and Carpenter set off for India in September of
1938 to go about what he called "the very nervy business" of trapping
an appropriate number of healthy rhesus macaques. He had to pay
off racketeering traffickers until enough animals were collected from
seven different provinces and brought to Calcutta. It took time and
further bribes to find a ship captain willing to take Carpenter and the
macaques home, "because 500 animals practically cover the whole
deck of a large freighter," he noted. When they finally embarked on
the USS *Coamo*, war had already made the Suez Canal unsafe, so
the ship made a test run around South Africa's Cape of Good Hope.
After further stops in Boston and New York, Carpenter and his
charges finally arrived in Puerto Rico in early December, forty-seven
days after leaving Calcutta. They had covered fourteen thousand
miles and Carpenter had worked fourteen to fifteen hours a day. "I
cleaned cages or fed animals all day long, exhausted, in rough weather
or calm," he said later. "You can imagine how delighted I was to turn
this shipment over."[38]

Though Cayo was established as a breeding colony, Carpenter had
a secondary goal: to create a nearly natural living environment for the
animals where it would be relatively easy to observe the monkeys' interactions. "I was interested in . . . the way a population organizes itself,
the kinds of social behavior that occurs," he said. His essential question
was a relatively simple one: Why do animals live in groups? But in 1938
social behavior wasn't high on science's list of priorities. The idea of

Cayo Santiago as a place to study social interaction was tucked inside the establishment of the breeding colony like money for health care in a highway appropriations bill.

Carpenter's expectation that, once released, the macaques might organize themselves in groups according to the regions they had come from proved a bad misjudgment. They mixed themselves up and a bloodbath ensued as the territorial monkeys tried to establish a new hierarchy. Dozens of animals were killed, including infants. Many males were driven out to sea and drowned. Much depleted, the remaining animals ended up in five distinct groups. Every monkey that lives on Cayo today is descended from those animals.

Despite his extraordinary efforts, Carpenter had little to do with Cayo after the first year or two. Nevertheless, his early observations were farsighted. He had spotted the similarities between human and nonhuman primate behavior at a time when few did—ten years before Harlow's work was widely known—writing that in observing monkeys one might glimpse the embryonic beginnings of human motivation and behavior, "free from cultural veneers and far enough removed to avoid the well known [sic] errors involved in man's study of himself."[39]

After this promising start, things changed on Cayo. World War II soon consumed the United States and both funding and interest in the monkeys dried up. By the mid-1950s, the monkey population had dwindled to fewer than 150 macaques, all of them in lamentable condition. Things might have gone on like that until they all died off.

Instead, Stuart Altmann arrived. The colony had just received some new money from the National Institutes of Health and had stronger supporters at the University of Puerto Rico. Altmann catalogued the remaining population—trapping, weighing, tattooing, and measuring every monkey on the island. The facilities on the island also had to be rebuilt. It took months, but Altmann finally restored sufficient order to be able to conduct scientific work again. Here, too, he was starting almost from scratch. Very little was known about macaque behavior, and the most prominent previous description seemed to Altmann "to do the capabilities of these remarkable animals a grave injustice."[40]

Like an explorer mapping a new territory, Altmann first had to establish the bounds of successful field observation. Problem number one: the monkeys knew he was there. His solution was to sit quietly in a spot with a good view and wait them out. "If the monkeys moved toward or around me, nothing was done to avoid them," he wrote. "At close range, persistent, direct staring was avoided." Occasionally, as necessary, Altmann communicated directly with the monkeys in language they could understand. When "attack seemed imminent," he threatened back by stomping his foot. If he needed to approach the monkeys when they were in a wary mood, he occasionally threw food to them. And he sometimes softly clucked and smacked his lips just as the macaques did to signal friendliness. He decided that behavior could only be considered social if it affected someone else. He used levels of communication to determine where the lines between groups fell and to arrive at his definition of a "society," so that those that were regularly in touch, so to speak, were a society separate from those that rarely communicated. Beyond that, he said, "I left it to the monkeys to tell me what are the basic units of their social behavior."

A few months into the work, Wilson came to visit. The two days he spent on Cayo Santiago, he later wrote, "were a stunning revelation and an intellectual turning point."[41] Wilson knew almost nothing about macaque societies but was captivated by what he saw. "I was riveted by the sophisticated and often brutal world of dominance orders, alliances, kinship bonds, territorial disputes, threats and displays, and unnerving intrigues. I learned how to read the rank of a male from the way he walked, how to gauge magnitudes of fear, submission, and hostility from facial expression and body posture."

By day, the two rangy scientists roamed the island. In the evenings, they compared and contrasted the social behavior of ants and monkeys, and termites, too. Then one day, while Wilson and Altmann sat

on Small Cay watching the insects and the primates, all the thinking that they had been doing about the behavior of different creatures coalesced into one question: Could it be that a unifying theory
explained the behavior of all these species and the way they interacted,
and that this theory might even extend to people?[42]

It seemed a radical idea. At the time—still the 1950s—it was generally believed that human behavior was a product of upbringing, of environment—nurture, that is. To say nature might also be at work was a
"no-no," as Wilson would learn. "Most of the social scientists had already
come to an agreement," he said later, "that the human brain is a blank
slate; that human behavior, including social behavior, is determined by
the accidents of cultural evolution and by learning alone and that there
was no such thing for the most part as human nature; that instincts do
not exist except in the most basic, primitive manner. That was dogma."[43]

⌒

Sitting there on Cayo Santiago, the pieces of evidence that had been
swirling in the scientific atmosphere suddenly coalesced into a visible
pattern, a pattern that Wilson saw with a bang through the same tinted
glasses as the European ethologists. This new view required accept-
ing that the behavior of an ant had something in common with the
behavior of Aunt Nancy. To be more precise, it meant that while ants
cannot be said to have "friends," there was more to the social life of the
ant than had previously been appreciated, but also that perhaps—this
was the radical part—that there was a little less individual agency to
the life of Aunt Nancy than she and other humans liked to imagine.
The social things that people did, such as sharing a meal or a laugh,
arguing angrily or celebrating joyfully, bonding with an infant and then
bickering with that same child in adolescence, bore some relation to
the things that animals did. Human nature had much in common with
animal nature. Even morality, Wilson would argue later, was guided at
least in part by the body and the brain, specifically by the genes that

shaped the body and the brain. Hate, love, guilt, and fear, the very emotions consulted by philosophers pondering good and evil, originate in the hypothalamus and the limbic system of the brain, noted Wilson. "What, we are compelled to ask, made the hypothalamus and the limbic system? They evolved by natural selection."[44]

Genes, or nature, this new argument went, play a role in social behavior, in the way we interact with our families and with our friends—just as Lorenz and Tinbergen had argued that genes affect how an infant duckling bonds with its mother. "Human beings inherit a propensity to acquire behavior and social structures, a propensity that is shared by enough people to be called human nature," wrote Wilson. Traits that existed across cultures, for instance, included the bond between parents and children, suspicion of strangers, and the tendency to be most altruistic toward close kin.

This new division of ethology became known as sociobiology, but to really take root it required what Wilson called "the most important idea of all."[45] That came from a soft-spoken British biologist named William Hamilton. When he published his breakthrough paper,[46] Hamilton was still a graduate student. Bookish, and a terrible public speaker who once said he wouldn't understand his own ideas if he'd had to listen to himself,[47] Hamilton nevertheless had the kind of simple, elegant idea that makes all other scientists wish they had thought of it: kin selection.

Hamilton was inspired by three things.[48] He was bothered by the problem of altruism in Darwin's theory of natural selection; it needed an answer. He also had a working knowledge of social insects. And he was intrigued by the mathematics of kinship—the fraction of genes shared by parent and offspring (50 percent), grandchildren and nephews and nieces (25 percent), and great-grandchildren and cousins (12.5 percent). Traditionally, natural selection was thought to occur directly between generations—I pass along to my three sons the genes that my father and mother passed on to me, and I am motivated to help my children and ensure their survival because we are related. After doing thought experiments involving an insect that reproduces by cloning and therefore shares more than half its genes with its siblings, Hamilton realized that if you consider selection at the level of the gene instead of the entire organism, genes are shared more widely among more distant kin. It follows that if I cooperate with my cousin, I am still furthering the cause for a subset of my own genes, even if the cooperation comes at some direct cost to myself. In purely mathematical terms, the idea was summarized this way: I will act altruistically at risk to myself to save one child or sibling, or four nieces, or eight cousins. This framework took the idea of "fitness"—essentially, the combination of traits that make up an individual's chances of reproductive success and survival—and split it into two categories. Direct fitness concerned an organism's own survival and reproductive success. Indirect fitness concerned the success of others with the same genes. The two together were dubbed "inclusive fitness."

Seen this way, altruism finally began to make more sense. Others started to link the motivation to help with the larger motivation to connect. In 1966 evolutionary biologist George Williams put the new idea this way: "Simply stated, an individual who maximizes his friendships and minimizes his antagonisms will have an evolutionary advantage, and selection should favor those characters that promote the optimization of personal relationships."[49]

There was one other important piece of the puzzle still to come. Robert Trivers, another Harvard graduate student, was captivated by Hamilton's logic, but thought an explanation was still needed for altruism beyond kin to explain the instances in which someone jumps into the water to save a drowning stranger. He developed a model for what he called reciprocal altruism that could explain cooperation and altruism in more nuanced ways.[50] Reciprocal altruism is the idea that if you scratch my back, I will scratch yours at some later date. In his argument, Trivers cited the symbiotic relationship between cleaner fish and their hosts, the warning calls birds give to other birds, and the many examples of human altruism.

The bottom line of reciprocal altruism was that the cost to the giver had to be less than the benefit to the recipient. The sticking point, as Trivers saw it, was the problem of cheating. What's to stop the recipient from taking advantage? He argued that many emotions had evolved specifically to address this issue. His model predicted friendship and the emotion of liking and disliking. It predicted what he called moralistic aggression—our strong reaction to perceived injustice or unfairness. It predicted gratitude and sympathy, which he theorized evolved to reward reciprocal altruism. And that wasn't all. It predicted guilt, trust, and reputation (learning from others about who can be trusted to reciprocate and who can't). These traits can be learned, Trivers argued. They are subject to change depending on circumstances early in life. And they can be used to cement friendship. He concluded by acknowledging the complexity of the system he was proposing. It was a veritable kitchen sink of emotion and motivation. Perhaps, he suggested, that very complexity might explain the increase in brain size seen in hominids since the Pleistocene Era. It took a lot of brain power to keep track of all this emotion after all.

The arguments for sociobiology were popularized by Richard Dawkins in his 1976 book *The Selfish Gene*, which introduced to a wider public the idea that genes rather than whole organisms could be the vehicle for natural selection. And then there was Wilson's 1975 book *Sociobiology: The New Synthesis*. In nearly eight hundred pages, he laid out the importance of studying social behavior at the level of populations and made a strong case for the genetic basis of social behavior.

The first twenty-six chapters of the book constituted an encyclopedic review of social microorganisms and animals. "Consider for a moment termites and monkeys," Wilson wrote, as if reliving that day on Cayo Santiago. "Both are formed into cooperative groups that occupy territories."[51] The book's last chapter began more provocatively: "Let us now consider man in the free spirit of natural history, as though we were zoologists from another planet completing a catalog of social species on Earth." In a more hypothetical tone, Wilson pulled from social sciences to examine how the biological foundations of human behavior might help us better understand our nature: "By comparing man with other primate species, it might be possible to identify basic primate traits that lie beneath the surface and help to determine the configuration of man's higher social behavior."[52] He suggested that if these behaviors were carried in the genes, we would have to look at the good and the bad: "Cooperativeness toward groupmates might be coupled with aggressivity toward strangers, creativeness with a desire to own and dominate, athletic zeal with a tendency to violent response, and so on."

Critics found the argument that "genes make us do it," that our social behavior is to any extent preordained and not culturally transmitted, unacceptable. A furor erupted. The strongest criticism came from colleagues right down the hall at Harvard. Led by biologists Richard Lewontin and Stephen Jay Gould, a group calling itself the Sociobiology Study Group wrote a scathing open letter in the *New York Review of Books*.[53] In their view, Wilson's argument was both unsupported by science and dangerous. "Wilson joins the long parade of biological determinists whose work has served to buttress the institutions of their society by exonerating them from responsibility for social problems," they wrote, after citing eugenics and Nazi Germany. Wilson, who was politically liberal himself, knew nothing of his colleagues' incendiary letter until he read it in the paper. He became so unpopular for a time that a young woman poured a pitcher of water over his head at a scientific meeting in protest.

The controversy and the science behind it made the front page of the *New York Times* in 1975,[54] and two years later, the cover of *Time*. "Why You Do What You Do" read the magazine's cover line over a photograph of a man and a woman being controlled by strings like marionettes.[55] The article featured Hamilton, Trivers, Dawkins, and Wilson. "The concepts are startling—and disturbing . . . ," it began. "All human acts—even saving a stranger from drowning or donating a million dollars to the poor—may ultimately be selfish. Morality and justice, far from being the triumphant product of human progress, evolved from man's animal past, and are securely rooted in the genes."

The idea would prove more resilient than the controversy.

TWO

Building a Social Brain

Our social lives begin at birth. "Humans are born predisposed to care how they relate to others," anthropologist Sarah Blaffer Hrdy wrote.[1] But parts of that natural predisposition are raw and fragile, and they must be nurtured. To make a friend and then to be a friend is a complex undertaking. Young children have to grow into the job because connecting positively and consistently with their peers places considerable demands on emerging cognitive and emotional abilities. Yet succeed they must. In the first few years of life, establishing relationships with other children is a major milestone. Parents must help.

Before he could have his friend Christian, my son Jake had to have me—and his father, and some loving babysitters—but, in his case, mostly me. He and I spent our days together. For the first few weeks that meant that he lay on my chest in the living room of our London apartment as, bleary-eyed, I found British television shows like *Watercolour Challenge* unaccountably gripping. Later, after we had moved back to Brooklyn, we dug in the sandbox at the playground, put puzzles together, and chatted about the potatoes I was peeling for dinner. We spent our nights together, too, or at least it felt like it in the early months when we were up every few hours, nursing and rocking, nursing and rocking. In each of those interactions, even the exhausted ones, when I smiled at Jake and he eventually smiled back, when I talked to him and he eventually talked back, when I laughed and he eventually laughed back, and when I cried and he stared at me and tried to work out what was going on with Mommy, he was honing the early social skills on which his later friendships would depend. We were engaged in phase one of the construction of his social brain—detecting and recognizing social cues and learning to attach.

The particular bond between mothers and babies is the special province of mammals, and it likely evolved as early as 225 million years ago. Reptiles and fish lay eggs and generally leave their babies to their own devices. Birds have stronger parental instincts, protecting their eggs and tending to fledglings in the nest. But mammals nurse, creating a powerful physical connection. The very word "mammal" derives from mamma, or mammary gland, and signifies the feeding of infants with mother's milk.[2] That bond is a key to mammalian social life, setting us up early to experience the appeal—and necessity—of others.

Among mammals, and even among primates, however, humans extend childhood far longer than any other species. This is the result of a trade-off. "Our rate of brain development is grossly slowed down, which has the positive consequence that we're able to let post-natal experience shape our brains much more. Of course, it also means we're born more helpless than other primates," says psychologist Mark Johnson, an expert in developmental cognition at both the University of Cambridge and Birkbeck, University of London.[3]

It is as if human babies arrive in the world a little underdone—so much so that some pediatricians call the first three months of life the "fourth trimester." Nature has compromised. To be able to walk upright on two legs, the pelvises of human females narrowed. Forty or so weeks of pregnancy produce a baby that's at the upper limits of what a woman can pass through her birth canal. Upon arrival, those babies are entirely dependent little creatures capable of not much more than eating, sleeping, and crying. Even the most besotted parent can be forgiven for thinking a newborn isn't a very social creature. The give-and-take of a social relationship—the call-and-response of acting and reacting—is wildly lopsided at the beginning. It's all give on the parent's part.

Yet an enormous amount of social infrastructure is being laid from the start. The senses—vision, hearing, touch, smell, and taste—are conduits through which a baby takes in the details of her new environment and conveys them to her brain, which is preprogrammed with a preference for social interaction, like a computer that is loaded with software waiting for the appropriate commands. Faces, voices, and loving caresses will set the program running. "From these small beginnings spring all the highly discriminating and sophisticated systems that in later infancy and childhood—indeed for the rest of life—mediate attachment to particular figures," John Bowlby wrote.[4] And neuroscientists have only recently begun to glimpse more clearly how this process—the creation of the social brain—works.

〜

It's a blustery November day when I visit the Babylab at the Centre for Brain and Cognitive Development at Birkbeck, University of London. Perhaps fittingly for a relatively new field, the Centre is housed in a modern brick building that is very unlike the gracious Georgian townhouses and squares of Bloomsbury that surround it. Despite the wind and snow flurries outside, the reception room is warm and welcoming. Smiling giraffes and elephants are painted on the yellow walls, bright red couches and blue and yellow beanbag chairs are set around the room, and multicolored rubber play mats cover a sizable stretch of the floor, with baskets of toys in easy reach. This is where the researchers greet their regular visitors, most of whom are under two years of age.

Tiny Aurora is among the youngest, only one month old. Her mother, Georgina, hands Aurora over to graduate student Laura Pirazzoli, who gently lays the baby girl down on a mat. Aurora is in constant gentle motion, rhythmically kicking her feet and pumping her fists. She appears to consider Pirazzoli's face as the young scientist leans in close.

"You're taking part in groundbreaking research," Pirazzoli says soothingly.

Pirazzoli unsnaps Aurora's onesie in order to attach leads for an electrocardiogram test. Then she strokes Aurora's belly affectionately with her forefinger and wraps her in a lamb's-wool blanket to keep her warm for the trip downstairs to the lab.

"You're a superstar."

Aurora is here for a study on touch, but that's just one of the social senses that interests the researchers at the Babylab. They want to understand the formation of what is known as the "social brain," a concept that has emerged in recent years and which takes up a lot of neural real estate. "Social cognition, the business of being able to see, understand, interact with and think about other people is a huge part of what our minds and brains and our fundamental natures are all about," says neuroscientist Nancy Kanwisher of the Massachusetts Institute of Technology in one of her lectures.[5]

To understand the social brain, it helps to think in terms of maps. Just as there are various ways of mapping the United States, there are a variety of ways of describing the brain. Geographically, the four lobes of the brain—occipital, temporal, parietal, and frontal—are like three-dimensional versions of the Northeast, South, Midwest, and West. In terms of evolutionary history, the automatic, emotional, and cognitive "layers" of the brain are relatively ancient, middle-aged, and young, rather like the American colonies, Louisiana Purchase, and then the Western frontier. (It's not actually so neat and tidy as that, but this triune layering of the brain, first introduced by neuroscientist Paul MacLean in the 1970s, is conceptually useful.)[6] You can also consider the brain in terms of travel and connection. The networks that join various tightly linked brain regions are like the highways and byways that cross the whole of the country. Some routes see a lot of traffic—it is easy to get from New York to Washington, D.C., because so many people are making the same trip every day and multiple roads, train tracks, and flight plans are set up to convey them; it's rarer to travel from New York to Wyoming and therefore a little bit more arduous. The same is true in the brain. Routes that are heavily used allow for faster connections just like superhighways do. The brain has to work a little harder to do what's unusual. But ask it to send messages along the same route often enough—even a long distance—and they will get there faster and faster.

It is this last kind of map that is essential to understanding the social brain. Everything the brain does depends on circuits of neurons that work together. Although people sometimes refer to a "center" for this or that in the brain, that term is inaccurate. Rather, different areas of the brain are linked to other areas via neurons, and scientists use a constellation of connection verbs to describe the relationships: certain areas are "associated with" or "mediate" or "influence" specific activity. The social brain is a loose confederation of structures and circuits that are distinct yet overlap and interact.[7] So, while reward, communication, and our ability to imagine what others are thinking (also known as mentalizing) operate along different highway systems in the brain, it's as if they all stop in Chicago en route. Reward messages, for instance, travel through the ventral tegmental area and nucleus accumbens on their way to the prefrontal cortex, but they also involve the amygdala, a region that helps determine whether the event being processed was positive or negative. Mentalizing, on the other hand, involves the superior temporal sulcus (STS) and temporoparietal junction (TPJ)—both of which prove critical to social interaction—but also the amygdala and the prefrontal cortex. In this analogy, the amygdala is Chicago.

In the beginning of life—such as Aurora's first days—the first part of the social brain to come online involves sensory processing. That's what provides the ability to detect and recognize who or what is going to be important in your environment.

Despite considerable progress in brain imaging, the neural happenings

inside the heads of babies remain one of the last frontiers in neuroscience. Most brain imaging technology, such as functional magnetic resonance imaging, is not very baby-friendly—not impossible to use, but not
easy either. Unless they're sleeping, babies move too much to deliver
reliable signals, and they can't respond to instructions or verbal cues.
They are small and likely to be weighed down by heavy equipment or
frightened by it. An MRI machine is nearly as loud as a rock concert.

Now, a promising new technology allows us to peer noninvasively
into the brains of babies, and the Babylab at Birkbeck is doing pioneering work in this area. Before observing Pirazzoli, I went upstairs to visit
Sarah Lloyd-Fox, the resident expert. A slim blonde in her thirties with
two young children of her own, Lloyd-Fox happened to begin work
at Birkbeck just as they were exploring the possibilities of functional
near-infrared spectroscopy (fNIRS) in very young children. "There
was no infant research on this," she says. "We saw NIRS as a window
into the social brain, and into the first year of life."[8] Her dissertation
focused on adapting the technique for babies. NIRS makes it possible
to watch what goes on inside the brains of babies and toddlers when
they are awake and responding to what they see, hear, or touch—it
works whether they are watching videos, interacting with caregivers, or
even just playing with their own toes.

As the name indicates, near-infrared spectroscopy makes use of light.
If you ever held a flashlight to your chin during childhood sleepovers,
you created a spooky red glow, which emanated from the light reflect-
ing the blood under your skin. The more oxygen that blood is carrying,
the redder the glow. The amount of oxygen in blood changes according
to metabolic demand. When brain cells are active, there is an increase
in blood flow as the neurons call in the resources they need to function.
Oxygenated blood is bright red. As oxygen is taken out of the blood it
gets darker, appearing more blue or purple. While you can't see blood
oxygen levels inside the brain just by shining a flashlight on your head,
near-infrared light does the equivalent. It is invisible to the naked eye
but can travel through the skin, skull, and brain tissue.

NIRS does have limitations. Most significantly, the signal has a shallow reach. It reveals what happens only in the outer cortex, the wrinkly mass of gray and white matter that makes up the brain's surface layer. Fortunately, what's of interest to Lloyd-Fox is social behavior, and there's plenty of that to be investigated in the areas reachable by NIRS.

First, she had to convert the necessary headgear to fit small heads. A black headband made of lightweight rubber—rather like a heavy-duty swim cap that's open on the crown of the head—is wrapped around the forehead and temples and closed with Velcro. On the youngest babies, a chin strap helps to hold the equipment in place. Along the sides of the head, brightly colored round buttons represent "channels" that sit above parts of the brain that are of interest. The buttons serve as both flashlight and receptor for the light waves that are reflected back. Black cables a quarter-inch thick sweep up and back from each sensor and are collected behind a child's head like a bundle of gravity-defying braids. It looks uncomfortable but the ease with which the babies take to it suggests it is not.

With the equipment ready, Lloyd-Fox and her colleagues set out to build on the classic behavioral studies of infant development from the past. Some of those studies were performed by Mark Johnson, whose office is nearby. "He's the one who started this," Lloyd-Fox says.

Johnson didn't start with babies, however. He started with chickens. As a graduate student at Cambridge University in the early 1980s, Johnson studied imprinting, the behavior made famous by Konrad Lorenz. Johnson wanted to know more about the visual, auditory, and tactile cues that underpin the imprinting process in baby chicks.[9] "The general textbook view had been that chicks hatch out into the world and then they imprint onto the first thing they see. They develop a strong social bond and attachment toward that first thing," Johnson says. "We showed that's only part of the story." Following Lorenz's lead, Johnson and his colleagues mounted a red box on a rotating stand and succeeded in getting chicks to imprint on the box. But Johnson also mounted a stuffed hen on the rotating stand. The chicks imprinted on the stuffed hen, too. "Whenever a newly hatched chick actually had a choice between a mother hen and some other object, they always chose the mother hen." It took a while to work out what was occurring, but eventually, Johnson says, "it clicked that there was this preference going on."

But what was that preference for mother based on? Perhaps it had to do with what Johnson called the mother's "henfullness," the biological significance of being a hen.[10] Or perhaps it had to do with visual complexity—that the contours and feathers of her body were more interesting than other objects. To find out, Johnson and his colleagues created mix-and-match versions of stuffed hens. They took them apart and put them back together, jumbled up like Picasso paintings, with limbs jutting out in unexpected locations. The crucial factor turned out to be groups of features assembled in certain reliable patterns that looked like heads and necks. Even a head alone, missing the rest of its body, elicited a preference. It didn't have to be a chicken either. The heads of other species, like a gadwall duck or a polecat, won the baby chicks' attention, too. "They were particularly interested in the head and neck region . . . that was the crucial predisposition or the bias that got them to look toward the mother hen," says Johnson, "which then meant that they imprinted and developed a social bond toward mother hens."

Johnson and others suspected that something similar was happening with human babies when they first encountered their mothers. Babies come into the world seeing through a fog. Their ability to distinguish between contrasts is limited, and the discovery of that fact is what led to the explosion of baby products in black and white. They also have no depth perception. What they can do is focus on a face that's about seven to twenty inches away, perfectly positioned to be stared at while nursing.

What's in a face? Quite a bit. They matter more than any other visual stimulus you see in the human environment. For babies, finding and attaching to a caregiver is a matter of survival. For children and adults, accurately reading faces is a critical piece of a healthy social life. It's one of the skills that people with autism find difficult. Babies who are disinterested in eye contact at six months are more likely to be diagnosed with autism a few years later.[11] Faces are founts of social information revealing identity, age, gender, and a range of emotions. Over our lifetimes, we can recognize as many as ten thousand faces—the average is five thousand—and can generally tell them apart quite easily in spite of the fact that they all share very similar geometry, and many share the same coloring or features, like snub noses or round cheeks.[12] Of all the features in a face, the eyes carry the most information. Are they wide or narrowed? Are the pupils enlarged or shrunken to pinpoints? Is the gaze direct or averted? Such details tell you whether someone is jubilant or dejected, content or cautious. And the distinct whites of human eyes make it possible to follow someone else's gaze to engage in joint attention.

Think of how difficult it is not to show what you are feeling. That is because evolutionarily, the ability to show emotion serves a purpose. Emotions are a way of communicating. More than that, they "organize our minds"—sorting our perception into rewards and threats and everything in between. A series of studies across human cultures found that the same six emotions are present at birth: fear, joy, disgust, surprise, sadness, and interest. (These findings have recently been challenged by neuroscientist Lisa Feldman Barrett, author of *How Emotions Are Made*,[13] but are still widely accepted.) By the middle of the second year of life, children start to develop secondary emotions that depend on social interaction. They can't get there until they've matured enough to have a social partner whose response to their behavior matters to them. These secondary emotions are pride, shame, guilt, jealousy, and embarrassment. All show up on the face. Facial features are a palette from which we work to create our social lives. Emotions represent reactions to the environment, a way of signaling to others. Accurately reading faces might mean the difference, for a child, between recognizing a buddy and a bully; when we're adults, it's the difference between recognizing a guy in a bar who's looking for a fight and a guy in a bar who's looking for a date. In many ways, the study of social behavior is the study of emotion.

Babies may be able to express joy, disgust, and the rest of the fundamental emotions, but can they recognize them in others? What do babies know about faces? Are they born with a preference for them? Or do they learn that faces matter over time? This was the question that Mark Johnson set out to answer. He had come across a small, controversial, and, therefore, largely ignored study from 1975 by a trio of researchers: Carolyn Goren, Merrill Sarty, and Paul Wu from the University of Southern California.[14] The paper claimed that newborns would move their head and eyes to follow a face—or something that looked a lot like a face, with eyes, nose, and mouth marked on a circular paddle. Johnson decided to see whether he could replicate Goren's study with a little more rigor.

Fresh from graduate school and with the confidence of youth, he thought it would be quick and easy. It took nearly four years. Just getting access to enough newborns was logistically difficult. His team eventually set up an on-call system—like obstetricians—so that they could be on hand as soon as a new baby was due to arrive. Within an hour of birth, a newborn was placed on its back in an experimenter's lap. The experimenter was armed with three paddles—one was blank, one displayed a simple drawing of a face (eyes, eyebrows, nose, and mouth), and the third had the same facial features scrambled, with one eye at the top, one at the bottom, nose and mouth upside down, and eyebrows containing the nose like parentheses. Holding up one paddle at a time, the experimenter began directly above the baby's face about seven to ten inches away. Once the baby fixated on the paddle, the experimenter moved it in a slow arc to each side, measuring how many degrees each baby turned his or her eyes or head.

Johnson and his colleagues repeatedly found a distinct preference for the face over the scrambled or blank paddles. They concluded that within an hour of birth, babies "possess some specific information about the arrangement of particular features that compose a face." They also tested older babies to see how long this phenomenon lasts and found that the preference for faces disappears at one month and then reemerges later.[15]

Johnson speculated that there were two processes at work. The first was a system to ensure that an infant orients toward face-like patterns. Given a newborn's limited vision, that system is probably controlled by more primitive, subcortical parts of the brain to make certain that the baby learns to recognize the most critical object in its world, its primary caregiver. Subsequently, Johnson believes that once the brain begins to mature, other more sophisticated areas take over the work of perceiving the world. Even just three months of experience, of seeing, changes the brain.

Johnson still thinks that is about right. That first paper, published in 1991, is regarded as a landmark in the study of how babies learn and develop cognitive skills. It has been replicated more than twenty times, with only one laboratory failing to find a similar result.

That doesn't mean the lively debate over the degree to which face perception is innate or learned is entirely settled. There are scientists digging more deeply into the specifics. Kanwisher, of MIT, made her name identifying a part of the brain called the fusiform face area. Intrigued by a condition called prosopagnosia, the loss of the ability to recognize faces, Kanwisher set out to look for a special part of the brain dedicated to recognizing faces. And she found it by looking in her own brain first.[16] She lay in a functional magnetic resonance imaging machine and her team showed her pictures of faces and pictures of objects. One particular spot in her brain was especially responsive to faces over objects. Roughly the size of an olive, it is located on the bottom surface on the brain about an inch behind and below the ear. The location had been described by someone else, but Kanwisher put it on the brain map, dubbing it the fusiform face area in 1997.[17] She believes her evidence is conclusive that the area is functionally specific in adults for recognizing faces.

At the other end of the argument is Kanwisher's academic neighbor Margaret Livingstone of Harvard. Livingstone and her graduate students recently spent four years raising rhesus macaques in the laboratory. From birth, the monkeys were not exposed to a face—human or nonhuman. Their caregivers wore welding masks when they interacted with the young monkeys. (For the record, the animals had lots of attention, activity, and toys. They were well nourished, and lovingly bottle-fed as infants. The only thing their worlds lacked was faces.) Using fMRI, Livingstone showed in 2017 that if the monkeys had zero experience looking at faces, the parts of their brains that should have been dedicated to that task did not develop the skill.[18]

Thankfully, however, human babies are not raised by people wearing welding masks, and do, in fact, see faces from the beginning. In the 1990s and today, Mark Johnson's view represents the middle ground. And practically speaking, it tells us what we need to know. When it comes to faces, humans are predisposed to seek out the particular combination of features that signals a face—top-heavy, two over one—but from the instant of birth, experience matters.

Experiments with NIRS have emphasized that experience matters on the order of minutes, not months. Sarah Lloyd-Fox kept the first experiment simple.[19] She focused on a part of the brain, the superior temporal sulcus, known in adults to detect both motion and social signals. She brought thirty-six five-month-old infants into the lab, sat them on a parent's lap, and played them sixteen-second videos of a female actor playing peekaboo or itsy-bitsy spider. In between videos, the babies saw still photographs of cars, helicopters, and other vehicles that served as baseline images. The question was would the blood oxygen level in the STS rise—indicating increased activity—when the babies watched the actor (social) rather than the vehicles (nonsocial). It did. To be certain motion alone wasn't driving the response, Lloyd-Fox did a second experiment that included videos of mechanical toys spinning or machine cogs and pistons in motion. Again, the babies responded more to the actor than to the mechanical toys. By five months of age, this particular part of the brain was already specializing in sociality.

Building on that first finding, Lloyd-Fox, Johnson, and their colleagues have expanded their explorations from the first hours after birth through to children's second birthdays. Most of their visual studies are variations on the first. The peekaboo and itsy-bitsy spider videos and the clip of the mechanical spinning toy are in regular rotation in the lab. In newborns between one and four days old, responses in the STS became increasingly attentive nearly by the hour—so that a baby born yesterday morning shows a different response than a baby born yesterday evening. Overall, across the first thousand days of life, there is a steady and robust process of neural specialization in the finer points of being social.[20]

Those results include auditory parts of the brain, too. Voices are as critical as faces. Understanding language brings with it an ability to communicate and extend one's social world beyond primary caregivers. Hearing begins in the womb around the beginning of the third trimester. That's why so many mothers are encouraged to talk or sing to their babies in the womb. From the baby's perspective, such sounds are muted by the fluid in which they are floating, and the full range of hearing isn't yet active. In the weeks before birth, babies hear loud sounds, but miss quieter ones as well as fine-tuned details. What they hear, mainly, is prosody, the rhythm and intonation of language. What they hear, in particular, is their mother's voice. A fetus's heartbeat speeds up when its mother is speaking. And once they are out in the world, most newborns find the sounds they heard in utero—anything from Mom's favorite singer to her heartbeat—soothing. They also clearly prefer Mom's voice to that of a stranger.[21]

How do babies know what to listen to? Adult brains are specialized for listening to human voices over nonvoice sounds, like bells ringing or engines running. But do babies enter the world with a preference for voices similar to their preference for faces? Using NIRS, Lloyd-Fox tracked infants' brains as they became increasingly tuned to human voices, a process that's akin to getting faster at picking out station signals from the static of a radio dial. Lloyd-Fox looked for and found different reactions to voice sounds like yawning, crying, and laughing compared to nonvoice sounds that babies would be familiar with, such as rattles and running water. The results showed a region in the temporal cortex, the same lobe that processes language, is increasingly tuned to human voices between four and seven months of life.[22]

Of all the voices in the world, babies continue to show a preference for their mother's even as they grow up and that preference has significant ramifications. In 2016 auditory neuroscientist Daniel Abrams of Stanford University School of Medicine asked children around the age of ten to listen to their mothers and to unfamiliar women speaking nonsense words and watched what happened inside the children's brains.[23] Nonsense words are useful in this kind of research because they put the focus strictly on voices rather than the content of what is said. The signature pattern evoked in the brain by the mother's voice (its "neural fingerprint") predicted the children's social communication skills. Abrams thinks this line of work can provide a template for investigating autism and other conditions in which the perception of voices, even a mother's, might be impaired.

After my conversation with Lloyd-Fox, I headed to the lab with Aurora and Pirazzoli. The door sported a sign with a baby in oversized glasses holding a book and the instructions DO NOT ENTER. BABY SCIENCE IN PROGRESS. The small, soundproofed room contained a baby seat for Aurora in front of a screen, a chair for her mother, and a set of computers against the wall that keep track of everything that's being recorded: video, heart rate, and eye movement.

The kind of touch Pirazzoli is exploring—known as affective or emotional touch—is a relatively recent discovery and an important addition to what we know about the social brain.[24] Of all the senses, touch is the most developed in a newborn. Our skin is like an advance scout, making contact with the world and signaling the brain about what it has discovered. Neurons in the skin take in information about everything we contact through a variety of nerve fibers and sensory receptors. Different nerve fibers respond best to different kinds of touch. They play favorites. Some like to be pushed, for instance, and others like to be stretched. One class of fiber, A-beta, does most of the work of discriminating, and these fibers are all over the body, especially in the palms and fingertips. Because they are sheathed in a fatty insulation called myelin, they are able to conduct the nervous system's electrical messages rapidly. Speed is of the essence, especially, for example, if you step on a thumbtack. C fibers carry messages about itch and pain. They are touch fibers of a different kind. They are unmyelinated and carry information at a much more leisurely pace, up to fifty times slower than their neighbors.

But there is one subset of C fibers that appears to be essential for our social lives. Called C-tactile (CT) afferents, they were first discovered in cats in 1939, but mostly ignored. Then in 1990, a group of Swedish scientists found them in people. Using a brand-new technique called microneurography, they inserted a needle through the skin of adults to touch a single nerve cell and record its electrical activity. The first several thousand such recordings were done on the skin of the hand. Every nerve cell there was myelinated and sent signals almost instantaneously to the brain. But then a young PhD student named Håkan Olausson and his advisor tried the same technique on the hairy skin of the forearm. When they touched the skin, it took three-tenths of a second before the signal reached the electrode. "It was a clear delay," Olausson remembers, "that was completely new."[25] What might such nerve fibers be for? Olausson wondered whether they had something to do with emotion—not what you feel, but how you feel about it.

The scientists located a French-Canadian woman named Genette who, because of a rare nerve condition, had lost her myelinated A-beta fibers, but still had the unmyelinated kind. Genette didn't think she could feel anything. Olausson blocked her view of her arm and brushed it softly and slowly with a paintbrush. He asked her whether or not she had been brushed. Each time she was brushed, "she was able to say that something was going on in her skin and she described it as pleasant," Olausson says. "She was almost 100 percent correct." Then the scientists put Genette in the scanner. Touch usually activates an area across the top of the brain called the somatosensory cortex, which processes tactile information—but Genette's brain responded differently. There was activity not in the somatosensory cortex but in the insular cortex, an area that is associated with emotion.

Nearly twenty years of work have revealed a lot more about CT afferents, as they're known. The fibers prefer a slow, stroking touch, at a pretty specific rate of one to two inches per second, just about the speed at which mothers caress their newborn babies. And they respond best at the body's natural temperature. They are not found in the palms of the hands or the soles of the feet, where so much discriminative touch takes place. Instead, they are present only in hairy skin, on the parts of the body—arms and shoulders, for example— where we are likely to receive an affectionate pat or a hug. And when CT fibers are stimulated, people routinely describe the sensation as pleasant.[26]

The scientists who study affective touch are convinced that is not an accident. Francis McGlone, a neuroscientist at Liverpool John Moores University in England who is one of Olausson's main collaborators, admits to being something of an evangelist for affective touch, and there is a messianic feel to the way in which he speaks about it. McGlone is convinced that this sense plays a far more significant role in human behavior than it gets credit for, forging connections and increasing our chance of survival. "Affective touch is a potential way in to understanding the development of the normal social brain," he says. "It's giving the brain knowledge of me and you, and the emotional quality of gentle nurturing touch is a very important feeling that underpins a lot of social interaction." Half joking, he even likens affective touch to the long-theorized but only recently discovered particle that resolved some of the puzzles of physics: "It is the Higgs boson of the social brain."

If that is correct, then working out what babies feel at birth is a critical next step. Multiple studies show that babies and children find affective touch pleasant. The Goldilocks-like touch that targets CT fibers—not too slow or too fast, but just right—has been shown to decrease nine-month-olds' heart rates and increase their attention to the brush. Children at a variety of ages prefer it to other kinds of touch.

But it's fussy. At Birkbeck, Pirazzoli is trying to test the responses of very young babies to affective touch. With Aurora snugly in the baby seat, Pirazzoli runs a paintbrush like the one Olausson originally used on Genette over Aurora's leg at timed intervals and at different rates of speed. A research assistant at the computer monitor marks the EKG readout at the beginning and end of each brushing bout. Pirazzoli admits to some frustration. She has yet to analyze all her data, but so far, her babies are not showing reduced heart rates. Science often works that way, taking two steps forward and one step back.

Whatever is happening in their social brains, around three months of age (or the end of that "fourth trimester") babies in human cultures all around the world master two critical skills that help cement their social bonds with caregivers. Look into their eyes and they are able to look back at you, a skill known as "mutual gaze." And they begin to show off a social smile.[27] It's a wondrous moment when a baby makes her mouth move in response to a caregiver's smile. Hearts melt. Mothers and fathers fall more fully in love. Bonds are set. "Parents often describe babies at this point as 'fully human,'" says Mark Johnson. We feel that way because once babies are able to respond to us, they have truly entered the social world.

That is what is supposed to happen anyway. Some of the most powerful evidence for the importance of sociability in a child's life comes from the instances where it is missing, such as in people with severe autism. Struggles with social communication are a hallmark of the condition with a wide range in how that manifests. Some children have barely any language and are incapable of responding when spoken to.

Others attend mainstream schools and test well in language abilities though they often have subtler problems with communication. I have interviewed teenagers in this latter group who rarely make eye contact. They glance at you but then their eyes slide away instantly. Or they close their eyes while they talk. "I definitely miss social cues," one boy told me.

Lloyd-Fox brought infant siblings of autistic children to the lab to test their social brain responses in the first six months of life. In the five children who subsequently developed autism by the age of three, she looked back at their early testing results and found reduced responses to visual and auditory social stimuli.[28] This work is in a very early stage and five children is an exceedingly small sample, but Lloyd-Fox hopes to expand it. "We think that the way in which they attend to stimuli might be different and that then impacts on social cognition in a massive way," she says. "If you had a different strategy for how to attend to things, that would [affect] how well you can see other people corresponding with you and [then] having friendships and communicating."

⌐

Although my son Jacob and his friend Christian first met when they were one-year-olds, they weren't yet "friends." Toddlers that age still have an egocentric view of the world and limited social graces. But toddlers' parallel play—one vrooming a fire truck along the rug while the other drives a toy car in circles nearby—contains the rudimentary beginnings of something more. They glance at each other, noticing what the other child is up to, sometimes imitating, sometimes grabbing. They are not yet cognitively capable of fully differentiating themselves from others or taking someone else's perspective. But they're on their way. That emerging capacity is called theory of mind. It is the landmark social development of the toddler years that allows young children to have friends.

Theory of mind falls under the larger umbrella of empathy, which is a good place to start unpacking what it requires. Empathy is an old concept enjoying new attention. I mentioned earlier that in the eighteenth century, economist Adam Smith was among the first to name this emotion, calling it "fellow feeling"—the sensation that something you see happening to another person is happening to you as well. The Germans call it *Einfühlung*, meaning "feeling into." At its core, empathy is the capacity to share and understand what another individual is feeling, like getting butterflies in your stomach when you watch a tightrope walker try to cross Niagara Falls. Recently, scientists have devoted increasing effort to teasing out empathy's complexities and the biology that underlies it. Most have come to see empathy as multifaceted, and to believe that many of those facets—at the simpler end of the spectrum—are shared by other species. Primatologist Frans de Waal of Emory University, a pioneer in the field, describes empathy as "a Russian doll" with "simple mechanisms at its core and more complex mechanisms and perspective taking abilities as its outer layers."[29]

Seen this way, empathy has three main components.[30] Emotional empathy, sharing another's feelings and matching his or her behavioral states, is a biological response found in many different species that probably evolved in the context of parental care and group living, both of which mean being exposed to and responding to the emotional signals of others. Cognitive empathy, which is where perspective taking or theory of mind fall, is the capacity to think about and understand another's feelings. And empathic concern, or compassion, adds the motivation to do something about another's suffering. Taken together, these components are fundamental elements of our social lives. "People empathize because it's absolutely critical for forming close relationships or relating to people at all," says psychologist Jamil Zaki of Stanford University, whose lab is devoted to studying empathy.[31]

Empathy of any sort begins with something very fundamental: the ability to know where you end and another individual begins. It requires a sense of self. Another basic form of empathy is emotion contagion.

When we see someone yawn, we often yawn, too. (So, too, do chimps and bonobos.) When children see that someone else has bumped a foot and then hold their own foot, it's not a sign of self-obsession, but rather the beginnings of empathy: the connection of another person's experience to one's own. Such self-referential behaviors predict later prosocial behaviors and empathic concern. De Waal calls this the "body channel" of empathy.[32]

In the 1980s, in one of the first tests of how and when prosocial behavior emerges in young children, Carolyn Zahn-Waxler, who was working at the National Institute of Mental Health, followed twenty-seven children across their second year of life.[33] Mothers were trained to watch closely when their child witnessed another person's distress. At certain points in the study, the mothers were asked to pretend to be upset. Shortly after turning one year of age, more than half the children had made at least one prosocial response. That generally meant hugging or patting whoever was crying or upset. By the time their second birthday rolled around, the children had expanded their social repertoires considerably. All but one could perform prosocial behaviors, such as providing verbal comfort or advice ("You be okay?" or "Be careful"), sharing, helping (giving a bottle to a crying baby), protecting, or defending.

By three or four, children begin to develop cognitive empathy, the ability to take the perspective of another person and to consider that person's experience of the world as separate from their own. It's a skill that children continue to fine-tune as their brains mature. In fact, maturation of the mentalizing network tracks nearly exactly with the development of explicit theory of mind in preschoolers. The progression in behavior is interesting to watch unfold. Neuroscientist Rebecca Saxe of MIT presents young children with a problem. She shows them a pirate puppet named Ivan who loves cheese sandwiches and puts his cheese sandwich down on a chest. Then Ivan walks away. A wind whips up and blows Ivan's sandwich onto the grass. Then a second pirate puppet, Joshua, comes along with his own cheese sandwich and places it on the chest where Ivan's used to be. Saxe asks the children which sandwich Ivan will take—the one that is sitting where his ought to be, or the one that is now on the ground? A three-year-old does not understand the concept of false beliefs. He knows that Ivan's sandwich is on the ground and assumes that Ivan knows, too. A five-year-old, on the other hand, completely understands that Ivan will think his is the sandwich that is sitting on the chest, right where he left it. Even so, the five-year-old joins the three-year-old in believing that Ivan is mean for taking Joshua's sandwich. It's only at the age of seven that a child understands that Ivan can't be held responsible for what he didn't know. "The wind should get in trouble," says the seven-year-old.[34]

There is a related skill coming online in these early years. Psychologist Michael Tomasello, who for many years was codirector of the Max Planck Institute of Evolutionary Anthropology in Germany and recently moved to Duke, studies how and why young children develop the social and communicative skills they need to cooperate—and by extension to make friends. He has carried those studies over into chimpanzees, looking for commonalities as well as the point of departure in humans, the thing that makes us human, in other words. In Tomasello's view, the critical skill that young children soon learn is what he calls "shared intentionality," the ability to create something with others and the necessary underpinning of sophisticated cooperation. In children, he says, by the age of one, helping and cooperation come naturally. As they grow, cultural experience drives the maturing of cooperative skills. Chimpanzees, Tomasello notes, are as capable of helping as young children are at the earliest ages, but they do far less sharing and are incapable of informing anyone about anything.[35]

I don't remember the exact moment when Jake and Christian crossed over from parallel play to something richer, but I can pinpoint that moment for my middle son, Matthew, and his dear friend Nicholas. Our families vacation in the same small beach town. Matthew and Nicholas were both only one year old the summer we met. They were the younger siblings, present but not really participatory, when their mothers or brothers were together. But the next year, when Nicholas was two and a half and Matthew nearly three, there was a dramatic change. One night over their chicken tenders, the two toddlers began talking to each other. When they were done eating, we let the boys perch on an upturned dinghy on the beach a few feet from where we were sitting. They kept up an intense conversation. Eventually, I had to take Matthew in to the bathroom. He could barely contain his eagerness to return to his new friend. As I pulled up his pants he tore back outside shouting, "I'm coming Nicky. I'm coming!" Not only was he eager to return to his friend, but he seemed sure that Nicholas was wondering where he had gone and what was taking so long. At any rate, Matty was convinced his new friend was as excited about resuming their conversation as he was. It was the beginning of a long-term friendship.

THREE

Friendship under the Skin

As I sat down to write this chapter, I heard an urgent cry. "We need you now!" Matthew yelled. We were at our family farm in upstate New York and our friend and business partner, Dan, had been bitten by an insect of some sort. He appeared to be having a serious allergic reaction. I ran downstairs and found him sitting on the porch steps looking pale. His heart was racing, and he felt dizzy nearly to the point of fainting. I grabbed my car keys and we headed for the emergency room and a dose of steroids.

The connection between friendship and health doesn't get simpler than this. A friend is someone who will drive you to the hospital should you need to go. More than that, a friend might be there to notice that you need to go. It is not hard to see how that can make you healthier and even lengthen your life in an emergency.

Friends also make us healthier by encouraging good habits. My friend Stephanie and I meet at 6 a.m. several times a week to run together. When the alarm goes off and it is still dark, I would not haul myself out of bed if she weren't waiting. Our runs have the added benefit of being good for the mind as well as the body. It is free therapy—a regularly scheduled chance to talk. I think of it as physical maintenance and friendship maintenance.

The influence of friends can also pull in the opposite direction—luring us away from healthy behavior. We are far more likely to stay at a bar or a party for "just one more" if we are with a large group of friends. That was the result of a 2015 study that checked in with Swiss young adults via their cell phones every hour during a night out. The more friends were reported present, the more drinks an individual consumed per hour. (The effect was stronger for men than for women.)[1] And then there is peer pressure, which as we'll see, can be exerted by presence alone. Teenagers take more risks just knowing a friend is on hand.

The connection between health and social relationships, however, goes far beyond handy car keys, lacing up your sneakers, or running an open bar tab. Over the last few decades, evidence has piled up to show that our relationships, including friendships, affect our health at a much deeper level, tweaking not just our psychology and motivation but the function and structure of our organs and cells.

For centuries, the idea that physical health might be affected by something as external to the body as a friendship would have been considered crazy. Anyway, doctors and public health officials had bigger worries. Well into the twentieth century, infectious disease was the world's predominant killer. In 1900 in the United States, pneumonia, tuberculosis, and diarrhea and enteritis topped the charts of leading causes of death. Together with diphtheria, they accounted for one-third of all deaths. Notably, just over 30 percent of all deaths occurred among children under the age of five, who never got the chance to grow old and suffer bodily wear and tear.

The situation changed markedly over the course of the twentieth century. Improvements in sanitation and hygiene, the discovery of antibiotics, and the implementation of childhood vaccines sent the death rate from infectious disease plummeting. By 1997 less than 5 percent of all deaths were due to pneumonia, influenza, and HIV infection, and less than 2 percent of all deaths were among young children. That considerable achievement meant that by the 1950s and 1960s, the medical world began to shift its focus to another kind of killer: cancers and chronic conditions like heart disease that took over the dubious distinction of killing more Americans than anything else. At the end of the twentieth century, heart disease and cancer accounted for 54.7 percent of all deaths followed by stroke and chronic lung disease.[2]

Once doctors began looking more aggressively for the causes of chronic disease, environment and lifestyle started to register as potentially critical. How people treated their bodies and what they put into them was now worth investigating. Strong evidence of the link between smoking and lung cancer first appeared in the 1950s,[3] and of the importance of diet and exercise in the 1970s.[4] Relationships, however, exist in your social world outside of the body. They are not something you ingest. They do not contain chemicals, even though we sometimes speak of them as "toxic." They are not directly linked to blood pressure in the same way that jogging is. Nonetheless, a handful of researchers wondered about the physiological effects of social integration. At the very least, they reasoned, having supportive friends and family ought to encourage healthy behaviors or trips to the doctor. But it was social integration's opposite—isolation—that made the connection clear. It emerged as a potential killer almost as suddenly as an iceberg appearing out of the fog.

To properly assess the effects of environment and lifestyle, you need to look at a lot of people. And that's just what epidemiologists set out to do. Beginning in 1948, they began to investigate health in the context of community. "[Community is] like the symphony orchestra," said Leonard Syme of the University of California, Berkeley. "You can study the violin or the trumpet or the drums to become an expert on the individual instruments, but that won't help you understand symphonic music."[5] Epidemiologists signed up adults in places like Framingham, Massachusetts (5,000 people), Tecumseh, Michigan (8,600), and Alameda County, California (nearly 7,000).[6] The communities in question were average places full of average Americans—Tecumseh was described as "neither particularly unique nor atypical."[7] Instead of looking backward and trying to account for what had brought on disease, these studies were prospective and longitudinal, designed to follow people as they aged and to look at the ways they actually lived their lives. Most included regular physical exams, although in some cases participants filled out health questionnaires. The goal was to tease out risk factors, primarily but not exclusively, for heart disease. It was the Framingham Heart Study, a project of the National Heart, Lung, and Blood Institute that is still running, that introduced the very term, "risk factor."

Along the way, in addition to recording physical measures, these studies began to probe their participants' social lives. Are you married? Do you belong to a church or other community organization? How many times in the average month do you see your family? Your friends?

A new notion about the potential power of relationships was in the air. It was rooted in two important ideas: stress and social support. The term "stress," as we use it today, was coined in 1936 by Hans Selye, an endocrinologist of Hungarian origin who did years of pioneering work at McGill University in Montreal.[8] Selye subjected laboratory animals to blaring light, deafening noise, and extremes of heat and cold and watched for subsequent changes in their bodies. While most of Selye's manufactured unpleasantness was physical, some was psychological—experiments designed to leave lab rats perpetually frustrated, for instance. Selye defined stress as the body's response to any demand for change. And he recognized that such a response could be adaptive. It seemed to be intended to help individuals cope with danger or move out of deficient environments. But in his animals, he also saw pathological changes, such as ulcers and swollen adrenal glands, and then eventually, if the stress persisted, heart attacks and strokes and other diseases very like those seen in humans.

We know now that when confronted with a stressful situation, all vertebrates—fish, birds, reptiles, as well as humans and other mammals—release hormones, such as adrenaline and glucocorticoids. Instantly, heart rates and energy levels go up. Selye was right: this is an evolutionarily ancient response designed to help us escape predators. When faced with a modern-day equivalent, such as a speeding car, the stress response helps you jump out of the way faster. Once the danger has passed, our systems dial back the signaling and slowly everything returns to normal. In an emergency, stress can save your life. And smaller amounts of it can motivate in more everyday ways. Stress helps us ace a test, nail a big presentation, or excel in a playoff game. The problem comes when stress never goes away, or when we anticipate constant challenges that never arrive. That's when we set ourselves up for disease. "Sustained or repeated stress can disrupt our bodies in seemingly endless ways," writes primatologist Robert Sapolsky, one of the foremost contemporary experts on the subject.

Even without the benefit of modern knowledge, Selye quickly made the leap from studying rats to worrying about what his findings might mean for people. Medical thinking at the time was still focused on infectious disease—bacteria or viruses that invaded the body and caused harm. Selye was proposing that the trials and tribulations of life could make a person sick—and could even possibly kill them. From there, it didn't take long for him to conclude that one of the most regular sources of stress for humans was other humans. "Stress in interpersonal relations," Selye later wrote, "is probably the most common problem we meet nowadays."[9]

Recognition of the potential importance of social support came later, in the mid-1970s. In one of those instances when intellectual lightning struck twice, two scientists working independently, epidemiologist John Cassel and psychiatrist Sidney Cobb, got the same idea at the same time: that social relationships might matter for health.[10] It was just a theory. Neither had proof. But both suspected that something about close relationships had a moderating, or buffering, effect on stress responses. They called that secret sauce "social support." Although it seemed like a new idea, the term was more like "really old wine in a new bottle," wrote sociologist James House. Implicit or explicit in literature, religion, and elsewhere, "it has just had different names: love, caring, friendship, a sense of community, and social integration."[11]

When I drove Dan to the hospital, that was social support. When Stephanie listens, as we jog around the park, to everything from my frustration over my boys' inability to hang up their towels to my sadness over my mother's dementia, that is social support. So is my husband's willingness to take on some of my household chores as I labor toward my book deadline. The concept covers a lot of ground. Emotional support makes us feel cared for. We employ it when we confide in someone or serve as a confidant, when we describe hopes and dreams, express anxiety, talk over problems, provide encouragement and feedback. Instrumental support is more tangible. It refers to pragmatic assistance, such as watching someone's child, sending over lasagna, carrying moving boxes, or lending money. And finally, there is informational support—instructions for treating the stomach flu or tips about sunny apartments for rent or promising job openings.

The first hard evidence of a link between health and social relationships appeared soon after the idea was proposed. It came from a young woman named Lisa Berkman, who was a graduate student in the social epidemiology program at the University of California, Berkeley, one of the first such programs in the nation. Berkman was interested in social integration. Before starting graduate school, she had worked as an outreach worker for a family planning clinic in San Francisco. The job entailed trying to identify people who needed help in each of the three communities the clinic served: Chinatown, North Beach (the Italian section of San Francisco), and the Tenderloin, which was very poor, full of single-room occupancy hotels and homeless people. The differences in the neighborhoods were stark. "It was so clear that some communities were so tightly integrated and cohesive and taking care of people. That was Chinatown and North Beach," Berkman says. "The Tenderloin was filled with people who were virtually completely socially isolated."[12] Many had arrived in the city by bus with only a few dollars in their pockets and fewer connections. Notably, their health was much more precarious.

When it came time to do her dissertation, Berkman, who now directs the Harvard Center for Population and Development Studies, wanted to answer the question her outreach work had produced: Did social isolation matter for health? She went looking for data and found the Alameda County study, one of those long-range longitudinal efforts established years before. Berkman and Leonard Syme, her advisor, looked at the level of social connectedness people had reported in 1965 when the study began and how many people had died by 1974. The results were striking: The fewer social ties people reported in 1965, the more likely they were to have died in the following nine years.[13] No matter how they analyzed the data, the effect never went away. Berkman and Syme showed that social isolation predicted mortality even after considering smoking, health behaviors, and socioeconomic conditions. This longitudinal association was an entirely new idea when they published their paper in 1979. It didn't prove causality but was provocative evidence that suggested a need for a wholesale shift in thinking. "I was surprised the results were as consistent and clear as they were," Berkman says. "It was an exciting moment." She wasn't the only one who thought so.

One recent summer day, I found Jim House among the books and papers piled up in his corner office at the University of Michigan's Institute for Social Research. He was on the verge of retirement, which explained the piles. He was culling the work of a lifetime. But he still possesses the strong intellectual curiosity he had as a graduate student at Michigan in the late 1960s and when he returned as a professor in 1978.

For anyone seeking to study relationships, the late 1970s and early 1980s were glory days at the Institute for Social Research. Sociologists, psychologists, economists, and public health experts flocked there to think seriously about how to appropriately measure social life. On the strength of his early work on stress and the workplace, House was already a pioneer in the field of social support, and the social psychological side of life was pretty much all that interested him then. When he read the Berkman and Syme paper suggesting that relationships had a strong, permanent, and physical effect on people, he remembers thinking, "Jeez, that's interesting."[14]

House went looking for the old data from Tecumseh, one of the other long-running studies that he knew well from his graduate school days and combined the study's measurements of social relationships with health information and mortality rates. Just as in Alameda, people who were more socially isolated were more likely to have died.

Soon, House was convinced that there was a bigger story to tell. He enlisted some junior colleagues and they pored through the existing (limited) literature, pulling out large-scale, representative population studies with relatively comparable follow-up data on mortality over six years or more. Only six such reports existed. Three were from the United States (Alameda County, Tecumseh, and one from Evans County, Georgia) and three were from Scandinavia. All had eerily similar results. "It was the kind of thing where if you went out and put this in front of [people], they would say you must have made the data up," House says. The degree of risk varied by study, but overall, social isolation roughly doubled the risk of mortality.

In the early talks House gave on the subject, he would use transparencies—PowerPoint hadn't been invented—and lay them down one after another beginning with Alameda, so that when he was done, he had created an echoing set of parallel lines showing a strong correlation between low social integration and higher mortality rates. It had the somber repetitive effect of a bell tolling, and it never failed to impress an audience. House knew that the editor of *Science*, one of the most prestigious and broadly read scientific journals, was looking to increase the magazine's coverage of social sciences and called him up. "I think I've got something."

The paper House and his colleagues Debra Umberson and Karl

Landis published in *Science* in July 1988 sounded the alarm about social relationships and health.[15] To provide a frame of reference, House had looked up some other known risk factors for health, such as smoking and obesity. When he compared death from all causes for smokers versus nonsmokers, and death from all causes for the socially isolated versus the socially integrated, he found that both had a two-to-one risk ratio. He did the same for other risk factors, and then he wrote a forceful conclusion: "These developments suggest that social relationships, or the relative lack thereof, constitute a major risk factor for health—rivaling the effects of well-established health risk factors such as cigarette smoking, blood pressure, blood lipids, obesity, and physical activity."

The results did suggest a chicken-and-egg problem, however. Did a lack of social relationships cause people to become ill and die? Or were unhealthy people less likely to establish and maintain social relationships? There was also a third possibility, that some other factor, such as a misanthropic personality, predisposed people both to have fewer friends *and* to become sicker faster. House and his colleagues went on the record on the side of the chicken. They argued that social relationships were probably driving health problems and not the other way around. What the paper did not say, because no one knew yet, was how or why social relationships exerted so much power. Were all kinds of relationships equally important? Did quantity or quality matter most? Was social support enough to explain the effects? These would be the questions for the coming decades.

The paper got people's attention. It made headlines across the country and ignited the field. And why wouldn't it? Declaring that having good friends and family in one's life was as important for health and longevity as quitting smoking was a public health version of Paul Revere's ride and House knew it. "That was Jim's stroke of brilliance," says Umberson, now a sociologist at the University of Texas, Austin. A twinkle appears in House's eye as he talks about it. "It did dramatize the thing."

Looking back, House recognizes that they made a bold claim, but thinks the congruence of the early evidence warranted it. And he has been proven right. A 2010 meta-analysis by psychologist Julianne Holt-Lunstad, which combined not six studies but 148 with a wider range of data on 308,000 people, found a 50 percent increased likelihood of survival for those with stronger social relationships.[16]

"After the fact, it seems kind of obvious and a little bit intuitive," House says. "So why the hell didn't we know and think about it earlier on?"

Among the many people inspired to start thinking about it was a social psychologist named John Cacioppo. At the time, Cacioppo was studying communication and persuasion, exploring how people formed and changed their attitudes about everything from disposable razors to health care. When he arrived at Ohio State University as a professor in 1989, Cacioppo also had an interest in psychophysiology, the use of physical measures as possible indicators of psychological states.

In 1890, William James suggested that feelings had to be connected to physiology. "No shade of emotion, however slight, should be without a bodily reverberation as unique, when taken in its totality, as is the mental mood itself."[17] But how to reliably measure and interpret such subtleties was another matter. For decades, the field was subject to overblown and conflicting conclusions and promises.[18]

One hundred years after James, the situation had not changed much. But Cacioppo was undaunted. An ambitious, whirling dervish of a thinker, he liked a challenge. He and his graduate school roommate painted a wall of their rented house in blackboard paint so that they could scribble ideas on it in chalk and debate their arguments through the night. "Working with John was like being shot out of a cannon," says a colleague.[19]

Cacioppo was convinced that scientists needed to merge their work on the body and the mind, and his interest in such questions coincided with a revolution. Neuroscience was transformed in the 1990s, the so-called decade of the brain, by the development of powerful new technologies. Magnetic resonance imaging (MRI), PET scans, and other imaging techniques meant it was finally possible to peer inside the skulls of living people and animals and try to make sense of brain anatomy and activity. But Cacioppo believed that looking only inside the brain wasn't enough. He called for studying questions on multiple levels—from the folds of the brain to the structure of society. "The brain does not exist in isolation but rather is a fundamental component of developing and aging individuals who themselves are mere actors in the larger theater of life," he and his colleague Gary Berntson, a neuroscientist, wrote in 1992. "This theater is undeniably social, beginning with prenatal care, mother–infant attachment, and early childhood experiences and ending with loneliness or social support and with familial or societal decisions about care for the elderly." Cacioppo and Berntson thought the inherent issues called for a new approach, a new field even—it called for a "social neuroscience."[20]

Both social psychologists and neuroscientists initially resisted the idea of social neuroscience on the grounds that the two fields had very little to say to each other—a resistance that helps to explain why nobody saw the connections between social relationships and health earlier. Nevertheless, Cacioppo the social psychologist and Berntson the neuroscientist teamed up (a collaboration that lasted until Cacioppo's untimely death at the age of sixty-six in early 2018).

Again, stress was a starting point. A colleague of Berntson's in immunology was studying herpes simplex in mice and looking for a way to consistently reactivate the virus according to the pattern seen in humans—the tendency for cold sores to break out just before the prom, for instance. The immunologist was trying various stressors, but nothing was working. He called Berntson, who suggested looking more carefully at people. "What are the stressors that typically reactivate [herpes] in humans? It's social status. That's where most of our stress comes from." The immunologist subjected his animals to the mouse version of prom: social stress in the form of isolation or the addition of a bully mouse into an established colony. "Lo and behold the herpes simplex reactivated," Berntson says. "It was clearly a social factor that triggered this difference in physiology."[21]

Meanwhile, Cacioppo had begun to work with a psychoneuroimmunologist named Janice Kiecolt-Glaser. Kiecolt-Glaser was interested in the physical toll on women who were looking after family members with dementia. She demonstrated, through punching small wounds into the women's forearms, that it took an average of nine days longer for caregivers' wounds to heal compared to unstressed controls (48.7 days versus 39.3). And there were signs of immune dysfunction in the caregivers that might have been to blame for the delayed healing, as well as differences in cardiovascular reactivity.[22]

Cacioppo's early collaborations with Kiecolt-Glaser revealed an important distinction, however. Not all caregivers were equally stressed. Those who felt they had social support did better. Women who felt stronger emotional bonds to the person they were tending showed fewer negative cardiovascular changes. Even that wasn't straightforward, however: if a couple had previously spent most of their time as a unit, socializing together and engaging in shared activities, the disruption to the caregiver's social life was greater and she showed more negative cardiovascular effects. Either way, it appeared that relationships were altering physical responses to stress and daily irritations.[23]

As the work progressed, Cacioppo began to think they needed to come at the problem from another direction—from the extreme end of the social integration scale. "Loneliness and the feeling of being unwanted is the most terrible poverty," Mother Teresa once said. Cacioppo wondered whether loneliness might be a most terrible health problem. Social isolation, the flip side of social integration, was the root of the risk that House, Berkman, and others had identified. Scientifically, the idea appealed for practical reasons. "There's something to measure," Berntson says. "When people are socially connected and they're pretty healthy, where do you go from there? But if you've got a specific pathology and you can trace it to social processes then you can get some sense for what the positive benefits of social contact are as well as the consequences of not achieving those benefits." When the John D. and Catherine T. MacArthur Foundation put together a team of researchers to dig more deeply into the broad question of social relations and health, Cacioppo convinced the group to focus on loneliness.[24]

The striking thing about loneliness is not that those who suffer from it are peculiar, but that they are so ordinary. Statistics on just how many people are lonely vary widely, but according to a 2018 online poll of twenty thousand people conducted by the insurance company Cigna, nearly half of Americans reported sometimes or always feeling alone. And a 2018 survey by AARP found that one-third of adults forty-five and older were lonely. (That result repeated their findings in the 2010 version of the study, but it should be stated that these surveys do not carry the weight of peer-reviewed studies.) Notably, the Cigna poll also found that the youngest adults, those between eighteen and twenty-two, had the highest levels of loneliness. An earlier study found that as many as 80 percent of people under eighteen reported being lonely occasionally.[25] As we see in chapter 4, it is likely that young people feel the pain of loneliness more acutely. I remember being fourteen and riding a commuter train from my home in center city Philadelphia out to my new high school in the suburbs where I knew almost no one and had been made to feel like an outsider. ("You live in the city? What does your father do?") As I stared out the train window, the lyrics to the recent hit "All by Myself" played in my head and I started to cry. It is a schmaltzy memory full of adolescent angst, but nonetheless makes me ache a little for that younger version of myself. At the other end of life, despite the popular impression to the contrary, the AARP survey found that loneliness decreased with age. (More on this in chapter 10.)

Although loneliness would occupy Cacioppo for the rest of his life, he regularly acknowledged that it was not his true subject. "The point in studying these isolated humans . . . is not an interest in loneliness per se," he said once. "It's an interest in the salutary social connections

that constitute what it means to be fully human. . . . Other individuals are, in fact, some of our greatest joys."[26] Cacioppo's colleague
Louise Hawkley, who worked with him for twenty years, agrees.
"Loneliness," she says, "was a convenient, helpful way of assessing a
fundamental need of human existence, that we are bred to connectedness, we are bred in connectedness, that we cannot avoid it and
that to the extent we have those needs thwarted we are going to pay
the price."[27]

But what does it mean to be lonely? For a long time, loneliness was
mostly discussed by philosophers, who tended to celebrate it as a necessary step on the way to making great art or achieving true reflection.
The dictionary definition is to be without companions, to be isolated.
But physical isolation and emotional isolation are not the same thing.
As German theologian Paul Tillich noted, there are two sides to being
alone. "Our language . . . has created the word 'loneliness' to express
the pain of being alone," he wrote, "and it has created the word 'solitude' to express the glory of being alone."[28]

Dutch sociologist Jenny de Jong Gierveld, a European pioneer in
loneliness research, found loneliness to be earthlier and more egalitarian than the philosophers and theologians who came before her. It
could be quantitative, as in an absence of social contact. And it could
be qualitative, when the problem is a lack of emotional closeness.[29]
Some people suffer from both at the same time, but it is possible to be
lonely in more nuanced ways. One can be lonely in a crowd, or in specific kinds of situations.[30] As writer Anaïs Nin put it, "Some people fill
the gaps and others emphasize my loneliness." For psychologists like
Cacioppo, the most important distinction was one of perception. Loneliness is how you feel about your social life. If you are truly introverted
and prefer to be alone, you are not necessarily lonely. But if you yearn
for connections you do not have, that is loneliness. It is the opposite of
feeling a sense of belonging or embeddedness, the opposite of feeling
one's intimate needs are met. According to Cacioppo, "the deeply disruptive hurt" of loneliness equals perceived isolation.[31]

When he first decided to investigate loneliness in the mid-1990s, Cacioppo began by having students at Ohio State fill out something called the UCLA Loneliness Scale, which had been developed in the late 1970s. It asks twenty probing questions. How often do you feel that you are "in tune" with the people around you? How often do you feel that you lack companionship? How often do you feel shy? How often do you feel that there are people you can turn to? The scale allowed Cacioppo and his team to sort students into high, average, and low lonely groups. Then he armed the students with pagers to carry as they went about their everyday lives. The pagers beeped nine times a day for seven days. Each time, the subject had to report on what he was doing and with whom, how she was feeling, and other details about the experience in the moment. On one of the nine days, Cacioppo collected blood samples and hooked students up to burdensome machines—it was not yet the days of miniature technology—to capture the workings of the cardiovascular system.[32]

The result was the first objective clue that physiological functioning differed according to levels of loneliness. The loneliest young adults showed differences in a measure of cardiovascular function called total peripheral resistance (TPR), one of the primary determinants of blood pressure. As the arteries constrict, resistance increases; as they dilate, resistance decreases. TPR was considerably higher in the lonelier students. Because they were young and their bodies had other ways of compensating, the students were still healthy, but their condition suggested trouble ahead. "It's the kind of pattern of physiological regulation that you'd expect would eventually lead them to have higher blood pressure," Hawkley says.

When Cacioppo and Hawkley moved to the University of Chicago (he in 1999, she soon after), they were able to approach the problem longitudinally. They recruited 229 older adults for a study that ran for most of the next decade.[33] Each participant underwent thorough annual physical exams, filled out the UCLA Loneliness Scale, and

joined in a lengthy social network interview. The latter required identifying up to eight or nine people to whom the person felt close, then
answering extensive questions about each friend or family member.
Age? Sex? How often do you see each other? How healthy is this person? How much do you like this person? Does person A know person
B? How about person C and person D? "[We'd] get the whole lay of
the land, a very rich measure of their social network," Hawkley says.
One of their first findings was that, just as they had suspected, those
early warning signs about loneliness and blood pressure in the younger
students did spell trouble in older adults. Lonely older people had significantly higher blood pressure than those who were less lonely.

But was loneliness causing these changes in physiology or was it
coming along for the ride? Perhaps another clinical disorder was giving
rise to both the physiology and the loneliness. Depression, in particular,
was regularly raised as a possible confounding variable with loneliness.
"Many people were saying if you're depressed of course you're going
to get lonely. You're just not going to be interacting, you're going to be
a downer to be around and so on," Hawkley says. She and Cacioppo
proved it could work the other way. "We could see that increases in
loneliness over one-year periods were causing increases in depressive
symptomology." And they were ultimately able to show that changing
loneliness—making someone less lonely—could, in fact, change depression. It was hard evidence that loneliness had costs worth addressing.

By the mid-2000s, Cacioppo and his colleagues had established
a frightening list of health factors linked to loneliness. In addition to increased mortality and depression, they showed evidence of
social withdrawal, decreased quality of sleep, elevated blood pressure, increased aggressiveness, and increased stress activity.[34] Other
researchers had found strong signs that more extensive social networks
protected against cognitive decline and dementia, and vice versa.[35]

To explain why loneliness had such power, Cacioppo turned to the
brain. He had long suspected that social support alone could not fully
explain the health effects of social integration or isolation. "The brain is

the organ forming, evaluating, monitoring, maintaining, repairing, and replacing salutary connections, as well as regulating the physiological responses that contribute to healthy lives or morbidity and mortality," he said.[36]

He developed a provocative theory: that loneliness begins as an adaptive response. Just as hunger signals us that we need to eat, perhaps loneliness was a warning bell designed to make us want to be with others, the social equivalent of physical pain, hunger, and thirst. "Evolution fashioned us not only to feel good when connected but to feel secure. The vitally important corollary is that evolution shaped us not only to feel bad in isolation but to feel insecure, as in physically threatened," he wrote.[37] His theory, in other words, was that perceived isolation makes us feel unsafe. "When you perceive that you're on the social perimeter, it doesn't just make you feel sad," he said, "it's also a threat to your survival. [Your] brain goes into self-preservation mode." In that state, like an animal on the edge of the herd, you become more worried about yourself than others around you, and you are hyper-vigilant to possible social threats. For many of us, these feelings serve as a reminder to call a friend. But unfortunately, for others who are chronically lonely, this state of perceived threat often serves to make the situation worse by weakening social skills. "The essence of social skills is you're taking the perspective of and [being] empathetic with others," Cacioppo said. That becomes increasingly difficult to do when you are lonely.

In the laboratory, Cacioppo and his colleagues explored this idea by inducing loneliness in more college students. Using hypnosis and prepared scripts, they altered people's emotional states by leading participants through moments in their lives when they had experienced both profound social connectedness and profound loneliness. They then administered psychological tests and found that even gregarious types showed poorer social skills once they were made to feel isolated. In one experiment, subjects tackled the task of identifying the color in which a string of letters forming a word was presented. Lonely subjects were slower to name the color associated with social words, such as "compete," and even slower if the social word had negative emotional associations, such as "reject."

"The delay indicated an interference effect," Cacioppo wrote. "Even when the task had nothing to do with sociality, and with no awareness of any intention to do so, the lonely participants were scanning for, and being distracted by, negative social information."[38] When you most need social connection, it would seem, you are the least able to achieve it. Using functional MRI, Cacioppo also found that the brains of lonely individuals show greater activation than the nonlonely to negative social cues over negative nonsocial cues. This "loneliness loop," Cacioppo believed, activates neurobiological and behavioral mechanisms that contribute to poor health outcomes.

Cacioppo's work makes the dangers of perceived social isolation clear, but what exactly is good about social relationships? What about them is protective—if indeed anything is? Is it quantity or quality? Is there a magic number? Do only the closest bonds have an effect? Does it have to be family? Or can friends fill in? What about variety? These were the questions that informed the next round of research.

The diversity of your social world is rather like the social equivalent of ice cream. Some people only ever choose between chocolate and vanilla while others leap at the chance to sample mint chocolate chip and sea salt caramel. In 1998 a psychologist named Sheldon Cohen at Carnegie Mellon University in Pittsburgh set up a complicated—and compelling—test of the importance for health of diversity in relationships. He convinced 276 volunteers to sacrifice themselves for science. They checked into a hotel, received nasal drops filled with cold virus, and then were quarantined for nearly a week. All participants also listed whether they were in touch, over an average two-week period, with anyone from twelve different possible kinds of relationships. For those of you playing along at home, they were spouse, parents, parents-in-law, children, other close family members, close neighbors, friends, coworkers, schoolmates, fellow volunteers, members of groups without religious affiliation, and members of religious groups. At the extremes, you can imagine a writer who is an only child, doesn't marry, and works at home (no siblings, no spouse, no in-laws, few interactions with colleagues) versus an Irish Catholic lawyer with multiple brothers and sisters who works in a large firm and has three kids and a husband (siblings, colleagues, kids, spouse, in-laws). Both can and usually do have friends, neighbors, and acquaintances from other activities, but the lawyer starts out ahead of the writer on the basis of pure arithmetic.

Diversity proved protective. Those with more types of social ties were less susceptible to common colds and did better on several other measures related to fighting off the virus. Those who reported regularly engaging with only one to three types of social ties were four times more likely to get sick than those who spoke regularly with people in six or more types of relationships.[39] That result held even when Cohen and his colleagues controlled for all kinds of possible confounding variables. It couldn't fully be explained by health behaviors such as smoking, alcohol consumption, and exercise, and it couldn't be explained by personality.

Others have suggested that a sense of control is what carries the most weight. In the influential Whitehall Studies, British epidemiologist and doctor Sir Michael Marmot found that social status was strongly correlated with health. This did not just mean, as most people assume, that the rich are healthy and the poor most likely to get sick and die. No. What Marmot found that was so intriguing was that at every step along the way, those higher up tended to be healthier than those just below them, and sicker than those just above. "Health follows a social gradient," Marmot wrote. He proved this by studying British civil servants working in a part of London called Whitehall. His work showed that access to care and resources to pay for it could not explain differences in health, but something else did. "Autonomy—how much control you have over your life—and the opportunities you have for full social engagement and participating are crucial for health, well-being, and longevity," he wrote in his 2004 book *The Status Syndrome*.[40] So, while his work highlights inequalities, the critical factors that underlie the disparities are degrees of control and participation.

And then, there is the question of quality. Relationships are not all positive. People criticize us. They let us down. They aren't always there for us when we need them, or alternatively, they refuse to leave us alone. As soon as scientists started peering more closely at the effects of relationships, it was logical that they would start start thinking about the quality of our bonds and whether it mattered. Social psychologist Bert Uchino, who was a graduate student in Cacioppo's lab at Ohio State and worked closely on the studies of Alzheimer's caregivers with Janice Kiecolt-Glaser, has devoted much of his career to this question.

Entirely positive ties—those that make us feel unconditionally supported—make up about half of our social networks and they have proven pretty easy to interpret. Studies show that these happy relationships provide a sense of purpose and meaning, they improve our outlook, and they encourage us to engage in better health behaviors. "All of those things we know are associated with better biological aging," Uchino, who is now at the University of Utah, says.[41] He has tracked physiological measures, such as levels of inflammation, levels of cellular aging, coronary artery calcification, and ambulatory blood pressure. To take one example, people who perceive high support in their lives showed less cardiovascular aging than people who didn't—to a dramatic degree. "They had blood pressure that was pretty close to someone who was 30 or 40 years younger," says Uchino. He found something similar when he looked at cellular aging. Each time a cell replenishes itself, the protective caps on the end of each DNA strand, called telomeres, shorten just a little bit more until eventually they can no longer do their job. The length of telomeres provides a handy clock by which to tell biological age as opposed to chronological age. In people with good relationships, telomeres were longer. "The link between biology and relationships is undeniable," Uchino says. We still can't say for sure that relationships are causing all of these positive health outcomes, but he thinks it's probable. "The reverse explanation would have to have something to do with how people who are biologically aging have poorer relationships," he says. "I guess that's plausible but it's less likely."

Purely negative relationships are rare but are clearly unhealthy. Uchino labels them "aversive," as in something we are driven to avoid, like milk that has turned. Simply put, they make us feel bad—mentally and physically. They are associated with stronger stress responses, higher levels of depression, elevated blood pressure, and reduced immune function.

So far so predictable—positive is good, negative is bad. Where Uchino has set out into new territory is in his work on ambivalence and the relatives, friends, and colleagues about whom we have mixed emotions. Positive and negative feelings are not always reciprocal. One does not have to go up if the other goes down; they can exist independently, as they do for that friend you have known for decades but find draining. These ambivalent ties make up a sizable part of our social world—almost half. In one study, Uchino even found that 80 percent of the subjects viewed their spouses ambivalently, although the more consistent figure for spouses is 50 percent.

In the cases of spouses and relatives, of course, these are institutionalized relationships that often bear more weight than friendship. Such ties are not easy to break, nor would you necessarily want to. Comedian George Burns captured that problem when he joked, "Happiness is having a large, loving, caring, close-knit family in another city." Friendships are easier to shed, but that doesn't mean we always do or always should. When asked why they keep up ambivalent relationships, most people say there is more good than bad in these bonds. Or they point to the pull of shared history, which acts like an anchor keeping them from moving on.

Fair enough, but Uchino's work is eye-opening. Since 2001, when he published his first report on the subject, he has been investigating how ambivalent relationships affect our health. At first, he thought they could go either way. The good might outweigh the bad. But it was also possible that the unpredictability of ambivalent relationships would give weight to the negative. He now thinks the evidence is unequivocal: ambivalent relationships are bad for us. "We have never found a case where ambivalent ties look beneficial for people in terms of their biology," he says. Across a variety of physiological measures—levels of inflammation, levels of cellular aging as indexed by telomeres, coronary artery calcification, ambulatory blood pressure—there was no sign that the positive parts of ambivalent relationships are helping anyone's health. Instead, the negatives take a toll. This holds true for intimate relationships, such as spouses and close friends, and for more distant ties, like colleagues, neighbors, and extended family members. Another nuance is that the perception of support is more helpful than actually receiving support, which can lead to shame or resentment or any of a number of other complicated emotions.

A caveat: very few other researchers are studying this question in the same way. Uchino acknowledges that his lab's calculations of ambivalence may exaggerate the number of relationships generating mixed emotions. People are asked to consider things like how helpful someone is when they need support or how upsetting the person is when they need understanding. Answers fall on a scale from 1 for "not at all" to 5 for "extremely." Anyone who receives a 2 or higher in both positive and negative measures—in other words, anyone who rates stronger than "not at all"—is categorized as ambivalent.[42] No wonder that amounts to almost half of the average person's social circle! Uchino fears making the cutoff higher because he suspects that people are hesitant to admit negative feelings and that their answers already underrepresent their true feelings.

Even as evidence mounted that relationships are indeed linked to health, a major mystery remained. How exactly does it work? What is happening at the molecular level that could explain how caregiving or connection or even just a casual conversation could make its way inside the body? What, scientists asked, are the mechanisms, the underlying processes? What are the pathways in the body that relationships travel?

The answer, to the extent we have one, begins with the fact that, stripped to its essence, everything psychological is biological. We take in the world, including the words and actions of our friends and family, through our hearing, vision, and sense of touch. The nervous system responds by sending signals to the heart, lungs, and muscles as to what to do next. Our stress responses, our immune systems, the blood pumping to and from our hearts, and the neurotransmitters firing in our brains all potentially play a role.

Even in much simpler animals, social bonds have been shown to be important—further evidence for their biological importance. In fish, for instance, scientists at Stanford allowed a male and female to form a pair bond. Then they separated them and allowed the female to see her male either win or lose a fight with another male. If the male won, there was activity in areas of the female's brain associated with bonding; if he lost, there was activity in areas associated with stress.[43] It seems that even without the learning and intelligence that are such an important part of human culture, something fundamentally social is going on— something that distinguishes closely bonded individuals from others.

Early on, Lisa Berkman, who had worked on the Alameda data, and her colleague Teresa Seeman acknowledged the mystery when they wrote, "Something about human relationships influences physical health and especially longevity." They proposed a bold idea: perhaps social relationships influence the body's rate of aging. As you grow older, the wrinkles and gray hair visible on the outside are mirrored inside by changes in organ function and cell structure. Arteries narrow. Blood pressure rises. The kidneys, lungs, and immune system all slow down. Our very cells age. It could be, wrote Berkman and Seeman, that social isolation or lack of support is chronically stressful and that organisms respond by aging faster.[44] Bert Uchino's work suggests they were on to something.

Others have focused on the immune system as a probable suspect.

Around about 2005, John Cacioppo attended a think tank meeting put on by the MacArthur Foundation. There he met Steve Cole from the University of California, Los Angeles.[45] Cole was interested in social effects on gene expression. Although codes for traits and behavior are written in our genes, they can be turned on or off by circumstance so that the mere presence of a gene is not always definitive. Genes sometimes stay silent, like an opinion that is never voiced. "For a gene to have an influence on what a cell does or how a human works, it needs to be expressed," Cole says. He believed that the social environment might have the power to determine which genes speak up and which do not. "On a molecular level, our bodies are artifacts, things that culture makes," he once told an audience in Los Angeles. "They are far more fluid and permeable to environmental influence than we appreciate."[46]

The process Cole studies is known as transcription. The DNA in our genomes describes a set of human possibilities, but it must be transcribed into RNA to make those possibilities a reality. When neurotransmitters or hormones are released by something we see or hear or touch, it sets off a chain reaction. Those chemical signals travel through the body and meet receptors on the surface of cells that are hooked up to on/off switches for gene expression within each cell known as transcription factors. The transcription factors then trigger the production of RNA and that RNA translates into proteins that go off and change cellular function and behavior. According to Cole, "that allows things that are fundamentally outside us, and may not even be physical events except perhaps at the level of photons falling on the retina or [sound] waves bouncing across my ear to become real biochemistry within the body that flips on or suppresses the activity of particular genes."

When Cole and Cacioppo met, Cole was presenting his research into why closeted gay men with AIDS were getting sicker and dying faster than openly gay men. At the time, the techniques of genomic analysis were far more restricted and only allowed for the measurement of one gene at a time. But if some men were dying faster than others, it had to be because the virus was replicating faster in those men. The genome of a virus—its full genetic complement—is simpler to analyze than the genome of a human being, and the genome of the HIV virus is especially simple, containing only nine genes. In the case of the closeted men, Cole found that the force of social stigma—a form of social stress—sped up the pathology of the virus.[47]

Intrigued by the implications, Cacioppo approached Cole after the presentation with questions. Might they investigate the effects of loneliness on human gene expression instead of the HIV genome? Might there be a genetic fingerprint of loneliness? And if so, could you find it?

Yes, Cole said, maybe. The entire human genome had been sequenced in 2003 and he believed that changed everything. "It allows us to analyze our way through the fog and confusion much faster," he told Cacioppo. "We can look for patterns—which groups of genes are doing what—that are very revealing."

They began by analyzing a small set of stored blood samples—just fourteen—from the elderly Chicagoans Cacioppo had been studying. The fourteen participants were at the high and low ends of the loneliness scale. That translated into the top and bottom 15 percent of the group. From that sample they created a list of genes in the white blood cells of the immune system, known as leukocytes, that were expressed differently in the lonely and the nonlonely.[48] The resulting graph resembles a political map of precinct-by-precinct voting in a swing state in that each spot changes color depending on whether or not the gene is expressed.

"It took about five minutes to make sense of what was going on," Cole says. "For anyone who was any kind of immunologist, you couldn't find a more clearly coherent set of genes." In the loneliest people, genes that governed inflammatory responses were up-regulated, or more likely to be expressed, and those that handled antiviral responses were down-regulated or less likely.

Cole was amazed. "It was uncanny," he says. "Why would all that stuff be happening with this supposedly trivial, annoying emotional state? Why would your leukocytes care about that?"

It wasn't just the stark result that convinced Cole that Cacioppo was on to something important with his studies of loneliness. It was the outpouring of emotion that greeted the paper they published in 2007. Nothing else he had done to that point got the same attention in the media or, more tellingly, from the general public. "We started getting emails from people in pain saying, 'I am so glad you are doing this because let me tell you this terrible story of this life I've led and how I got sick,' or 'this aunt of mine who can't feel close to anyone and later got an autoimmune disease.'" The message was clear to Cole: "This is what people really care about. It's sufficiently important to the real world that I have to keep following this up. These lonely people were really suffering, all the way down to their genes."

FOUR

Middle School Is about Lunch

arly in 2010, the year we moved to Hong Kong, our three boys were eleven, eight, and six. When we sat them down to tell them we'd be moving there for a few years, we tried to sell it as a challenging adventure. Their responses were mixed. Jake was anxious. Alex, our baby, was excited. But Matthew, our middle son, was inconsolable. He was having none of it.

"What about my friends!" he cried.

We tried to reassure him.

"You'll be back . . . you'll have your family . . . you'll make new friends."

Matthew looked at us with anguish and said, "And then I'll have to leave them, too."

It went on like that for weeks.

Matthew has always been the outgoing one. He seemed to recognize at a very early age the sustenance he got from other people. As a toddler he waved and grinned at strangers from his stroller. His second-grade teacher once told me he might have been the funniest child she ever taught. Another mother once marveled at his "profound social skills."

Such gregariousness generally gives him confidence. And it wasn't the prospect of making new friends that worried him. It was the wrench of separation from the ones he had. Matthew's friends felt like his whole world.

In this, he was age appropriate. The ability to make and keep even one close friend has been seen as vital to children's well-being for more than half a century. What has changed is that we now understand at a biological and even evolutionary level why that should be so. And we are beginning to appreciate that the intensity of feelings generated by friendship—or loneliness—in childhood and adolescence is by design.

I've already mentioned that a notable feature of human childhood is the sheer extent of it compared to every other species. We may not stay children forever, whisked away to Neverland by Peter Pan, but we experience an unusually long layover between weaning and puberty. The complexity of human brain development takes time. Much of that time is spent honing a new, more advanced set of social skills.

It begins with a period known as the five-to-seven shift, when children's cognitive abilities gain sophistication. In the theory established by Swiss psychologist Jean Piaget, this is the phase when children move beyond their concrete view of the world and become capable of logical reasoning (what Piaget termed "operational thinking"). A five-year-old, for example, might not understand that his mother could also be the sister of his aunt, while his seven-year-old sister understands perfectly. Other theorists like Sigmund Freud and Lev Vygotsky also singled out a cognitive turning point around this time. One of its hallmarks is the expansion of language skills, which bring a wealth of new social options.

Even outside of psychology, many cultures have historically described the "age of reason" as beginning around seven.[1] That is reflected in the traditional start of formal schooling at five or six, or the increased responsibility given to children in hunter-gatherer societies. As children move out of the orbit of the family and beyond their first caregivers, the web of their relationships widens. They meet other children and the fraction of social time spent with peers triples. Peer groups expand. Adults do less and less supervising of children's interactions. Kids begin to care more about how they are perceived by their friends. All of this is part of a necessary developmental process. Parents are the primary agents of socialization early on but then that role bumps up against some limits. Relationships with parents are too vertical and unequal to allow for all that children need to learn to do socially. Siblings help but come with their own rivalries. When kids arrive in school, they encounter horizontal relationships.[2] They must begin to learn about cooperation and collaboration, reciprocity, and loyalty and what it means to be trustworthy. Friendship is where they do that.

A three-year-old girl once described two classmates to me. She called them her "best friend" and her "worst friend." After quizzing her a bit, I realized that she didn't like the boy she called her worst friend at all. Presumably, she didn't know what else to call him. Preschoolers do know what's good about friends, though. Asked to define a friend, they reel off three observations: "A friend plays with you"; "A friend shares with you"; "A friend doesn't hit you."[3] Out of the mouths of babes, I tell you. Playing, sharing, and not hitting reflect young children's still very concrete thinking. These are recognizable, observable actions. All count as prosocial behavior, in that they are voluntary and intended to benefit others.

With time, children's expectations of friends become less physical and more relational and abstract, but the foundational bones are still there if you dig for them.[4] In adolescence, play turns into hanging around. Sharing turns into helping. Loyalty and intimacy become more central requirements, especially for best friends. Friends become important sources of validation and support, and children learn to provide support as well, something they are not generally in a position to do for their parents.

Even in the earliest years, temperament helps to determine how we socialize. Developmental psychologist Jerome Kagan found that 10–20 percent of babies begin life "inhibited." They show clear discomfort at anything new, kicking and crying readily. As those babies get older, they are more consistently shy and are more likely to be introverted, although some grow out of their social anxiety. (Kagan always stressed that this subset of people who were "high reactive" as babies may be less sociable, but they often are more attentive to the friends they do have and make careful scientists, thoughtful writers, and so on.) At the other end of the spectrum, about 30–40 percent of babies come into the world "uninhibited." Highly social from the beginning, they are bold and interested in new people and things. These babies tend to grow into more gregarious adolescents and adults.[5] If you've done the math, you've realized that this leaves roughly half of us somewhere in the middle. Children can also be measured along the same five personality factors as adults—openness to experience, conscientiousness, extroversion, agreeableness, and neuroticism—but the stability of these traits increases up to about age thirty, when they finally gel.

Gender also plays an undeniable role in the choice of friends. From preschool through adolescence, boys gravitate to boys and girls to girls. Indeed, friendship with opposite-sex peers "drops off precipitously after seven years of age."[6] Perhaps that is part of what drew Jake and Christian together. They were boys who liked similar things—playing with Legos and collecting Pokémon cards. Unlike Matthew, Jake and Christian were shy when they were young, but not around each other. They felt comfortable and confident in each other's presence. In declaring themselves best friends at four, they were like three-quarters of preschoolers who claim to have such a close buddy.

It is no accident that young children think of play when they think of friends. Playing is what they spend most of their time doing. It is also how they need to spend most of their time for developmental reasons. "Play is a profound biological process," wrote psychiatrist Stuart Brown, who devoted much of his career to convincing the world to take play seriously. "It has evolved over eons in many animal species to promote survival. It shapes the brain and makes animals smarter and more adaptable. In higher animals, it fosters empathy and makes possible complex social groups. For us, play lies at the core of creativity and creation."[7]

And not only for us. Dogs tug and fetch. Cats pounce and chase. Baby chimps wrestle and tickle. Fawns gambol. Rats play, too. After identifying ultrasonic rat chirps with a repurposed bat detector and labeling the results as laughter, neuroscientist Jaak Panksepp figured out how to induce more chirps by tickling his rats. "Lo and behold, it sounded like a playground!" he reported.[8] E. O. Wilson is convinced that ants engage in play fighting. In his book *Play*, Brown described watching young brown bears cavort in the wilderness. "The two bears went in and out of the rapids, splashed through clear sparkling pools, circled, pirouetted, then stood and leaned against each other, embracing in an upright dance. Periodically they paused, looked at the water, and then, as if under the influence of a master conductor, set at each other mouth-to-mouth, head-to-head, paw-to-paw, in an agile display of bear play."[9]

Those two bears lived in Alaska, where Brown was observing the work of psychologist Bob Fagen, one of the first scientists to see the value in studying animals at play. Fagen found that the bears that played the most, survived longest. "I believe that play teaches young animals to make sound judgments," Fagen told Brown. "For instance, play fighting may let a bear learn when it can trust another bear and, if things get too violent, when it needs to defend itself or flee. Play allows 'pretend' rehearsal for the challenges and ambiguities of life, a rehearsal in which life and death are not at stake."[10]

Even while watching bears, we usually feel that we know play when we see it. But in order to study something, scientists have to be able to define and measure it. While there's some variation still in definitions of play, most converge on a few essential points. Playful behavior resembles a serious adult behavior but is done by a young animal or is exaggerated or awkward or otherwise altered. It serves no immediate survival purpose and appears to be voluntary and pleasurable. Play also happens only when animals are not under stress and have nothing else to do. In other words, it's fun.[11]

Why is this ostensibly unnecessary activity so pervasive? Because play is not unnecessary at all. It is vital. The fact that we see it in so many other species—like so many of the social behaviors this book discusses—tells us that it serves a larger purpose. As does the fact that it's fun. The brain is designed to make us like things that are good for us. "We are wired to like activities that are helpful for our survival," Brown writes. Those activities trigger surges of dopamine, a signaling chemical generated in parts of the brain called the substantia nigra and the ventral tegmental area. In a chain reaction, the dopamine releases more chemicals that make us feel good: noradrenaline and adrenaline. The main stress hormone, cortisol, stays silent. If you find play to be a source of stress, you're not doing it right.

It is actually hard to deprive animals other than humans of play. If you manage that trick, however, the animals become aggressive. In people, a lack of play or abnormal play in childhood is often a sign of a neurological disease or disorder like psychopathy. Early in his research into play, Brown started taking "play histories" from patients. He studied prisoners on death row in Texas, asking them to describe how they entertained themselves in their childhoods. They didn't, he discovered. There was a notable absence of play.

Panksepp, who discovered play in rats, named play as one of seven primal emotions. "Play is a brain process that feels good, that allows the animal to engage fully with another animal. And if you understand the joy of play, I think you have the foundation of the nature of joy in general," he told *Discover*. "Play is engaging in an attachment-like way with strangers, which you have to do later in life."[12]

Another important landmark in play research came from neuroscientist Marian Diamond. In the 1960s at the University of California, Berkeley, Diamond was also studying rats. In order to work out what kind of early environment was healthiest, she set up experiments in which some of her rats had cages full of toys and other rats, while other animals had only the bare essentials. The rats raised in what Diamond called "enriched environments" were smarter, with bigger, more efficient brains. Access to play stimulated brain development. And it didn't just have to do with toys. "The combination of toys and friends was established early on as vital to qualifying the environment as 'enriched,'" Diamond said later.[13]

Clearly then, play accomplishes a lot. It allows young animals to learn without stress. It invites exploration. It fine-tunes self-control, both physically and emotionally. It serves as practice for things that will be done later in life. That's why so much animal play comes so close to more serious adult endeavors like hunting and fighting (think of kittens pouncing on a ball of yarn . . . or a mouse). And it is why children who engage in imaginary play often pretend to be adults—playing house, cops and robbers, or alas, war.

Mammals, the most social of creatures, are also practicing being sociable when they play. Zoologist John Byers of the University of Idaho counted up the amount of play in various species and measured it against the development of the frontal cortex, the area of the brain responsible for higher-order reasoning and planning. He found that play is most prevalent among species with big brains. And the period of maximum play in each species is tied to the rate and size of growth of the cerebellum. He speculates that play helps to sculpt the brain.[14] Furthermore, the more years a species requires to grow to maturity, and the longer it requires parental care, the more it plays. If a species has a big brain relative to body size, it probably engages in social play. Those with more neocortex typically live in larger social groups and have more complex social relations. It follows that they have more playmates.

Let's review. Species with long childhoods, big brains, and large social groups play more than any others and they do the lion's share of that playing during the time in life when their brains are growing fastest. Sound familiar? Play is the work human children put in to grow their brains and their social skills. Babies don't play with one another, but they do play with their own bodies, grabbing hands and toes. Once they can gaze at their parents and smile at them, they are nearly always up for a game of peekaboo, which we know from Sarah Lloyd-Fox's work engages the social areas of the brain. Within a few short years, human children develop a wider repertoire of play than animals, especially as they gain language. They can play with objects, like blocks or cars or dolls. They can play by leaping, running, and jumping. And they can engage in social play, the world of imagination and pretend that is such a charming feature of childhood. This last type of play, social play, is especially noticeable in animals that demonstrate a lot of change or plasticity in behavior over their lifetimes, and there is no young animal more plastic than the human child.

"Middle school is about lunch."

I turned and looked at Mary, the woman who had spoken. We were sitting on a beach watching our kids swim. It was August, and though we were still on vacation, our thoughts had turned to the coming school year. Jake would be starting middle school that September. Mary's oldest daughter was a few years ahead, so she was sharing her wisdom, which seemed profound to me.

As they reach middle school, children drift away from the pure play of running in the yard at recess or building with Legos. Middle school brings the beginnings of puberty for some, first crushes for many, and a shift from child to teenager for all. It brings higher levels of academics. But if you want to know whether your child is going to be happy or miserable, confident or anxious, being a fly on the wall at lunch would probably tell you a lot.

Initially, the biggest shift middle school brings is one of context. It is a change of environment that marks a milestone, the transition from elementary school. Most American students move from spending the bulk of the day in one classroom and with one set of classmates—a social bubble of sorts—to multiple classrooms and multiple new classmates. Their number of potential social possibilities swells. Friendships change their function in that shift. Peer relationships become much more nuanced. Children are entering a period of maximum concern over acceptance or rejection, over who does what to whom, and over how they will be perceived.[15]

No wonder lunch looms large. In many schools, it is the time in the day when these preteens have the most agency. It is why the movies are filled with so many scenes of anxious children holding a tray and not being sure where to sit. If we needed a reminder of the intense vulnerability lunch period brings, we got one in the efforts of a teenager named Denis Estimon. When he was a newly arrived Haitian immigrant in a Florida elementary school, lunch was the worst part of Estimon's day. He decided to do something about it when he reached high school and cofounded a club called We Dine Together. "It's not a good feeling, like you're by yourself. And that's something that I don't want anybody to go through," Estimon told CBS.[16] Club members spend the lunch hour wandering the cafeteria and courtyard of their Boca Raton school in search of anyone eating alone. Then they sit down with their own lunch and chat.

Out of curiosity, I spent a few minutes recently standing on a street corner outside my neighborhood's largest middle school, which allows students to leave the building for lunch. A group of three girls giggled and whispered together as they crossed the street to the deli. Two boys dribbled a basketball on the sidewalk as they headed to a nearby court. Everyone studiously avoided the playground in the park across the street—presumably they thought they had outgrown monkey bars. There seemed to be some early attempts at flirting going on inside the pizza parlor behind me. These were little kids turning into teenagers. The change was unfolding before my eyes like time-lapse photography.

Jaana Juvonen doesn't stand on street corners and watch middle schoolers interact in order to guess at the quality of their friendships. She asks them. Juvonen is a developmental psychologist at the University of California, Los Angeles. Appropriately, she and I met for lunch to talk about her work, although the café near the UCLA campus where we ate was full of adults, not middle schoolers.

About ten years ago, Juvonen set out to capture how peer relationships change over the course of adolescence. Over a period of three
years, she and her team recruited six thousand sixth-graders from
twenty-six different middle schools in Los Angeles and have been following each cohort ever since. When we met, the oldest students, the
first group to join the study, had finished their first year in college.
Each year, each participating child filled out a series of questions about
peers. Name your closest friends. Does this kid have your back? Can
you talk to him or her about anything? Do they come to your house?
Have you ever been bullied? Have you seen anyone else be bullied?[17]

The study revealed that instability rules, at least at the beginning.
Two-thirds of the sixth graders changed friends between the fall and
the spring of their first year of middle school. Juvonen suspects that
has to do with the structure of the school system. Students arrive from
smaller elementary schools knowing a few other children from fifth
grade. At the start of the year, they stay close physically and emotionally to those familiar classmates. But as they settle into life in their new
environment, sixth graders' social horizons expand. They gravitate to
those with similar interests of the kind that begin to solidify in these
years—soccer, theater, robotics. Similarities, as always, attract. Often
earlier friends fall by the wayside.

Friendship has real power for kids. Juvonen thinks that friendship
may even begin to resemble an attachment relationship like what
children have initially with parents. "[These] are really very, very
close and emotionally intimate relationships," Juvonen says. "And
even if that particular relationship doesn't last it has ramifications on
subsequent relationships."

Too often educators and parents fail to appreciate the potential
upside of these strong ties. Teachers often separate friends whose banter can be disruptive. Yet when researchers record student conversations during class, while kids are problem solving or working together,
there is evidence that students collaborate more effectively with their
friends. "Their dialogue is much deeper, cognitively more complex than

when we ask kids to work with just any classmate," Juvonen says. "It's really interesting that we as adults in the society often regard friendships more as a nuisance and a distraction rather than give them the value that they really deserve. They are incredibly important for kids."

But there is also a dark side to the social world of middle school, as anyone who has been through it will remember. Sixth graders who do not have friends are at risk of anxiety, depression, and low self-esteem. About 12 percent of the six thousand sixth graders in Juvonen's study were not named as a friend by anyone else. They had no one to sit with at lunch and no one to stick up for them when bullied. Of that group, boys outnumbered girls nearly two to one, and African American and Latino students were more likely to be friendless than white kids. Inspired by John Cacioppo and Louise Hawkley's work on perceived social isolation and the sense of threat that comes with it, Juvonen and her student Leah Lessard investigated whether perceptions of social threat could explain the mental health difficulties that beset friendless middle schoolers. Their hypothesis was that not having friends in sixth grade triggered a greater sense of threat in seventh grade, which led to increased internalizing difficulties, such as depression and anxiety by eighth grade. That is just what they found. It wasn't friendlessness alone that created problems, it was the resulting sense of threat.[18]

Then there is bullying, which Juvonen has studied extensively.[19] "Friendships take place in this larger context where there's a status hierarchy," she says. "In some schools or some settings, it's much more explicit than in others, but even in those settings where there's not one obvious hierarchy, there are certain distinctions. Kids know very well which kinds of kids are friends with one another and where they stand in that overall status hierarchy." Most of the time, bullying is a very strategic effort to gain and maintain status, she says. If anything has surprised her it is how consistently popular bullies are, at least in the short term. "Why should they change their behavior?" she now says. "The rewards are so great."

Juvonen and her colleagues have closely examined the role of friends for children who are bullied. As you might expect, those with the sparsest social networks were the most likely targets. But did it help or hurt, the researchers wondered, for a child to have a friend who had also been victimized? Hanging out with other victims could just make life worse, or it could make a child feel less alone. They found support for the latter.[20] "Shared plight helps," Juvonen says. And children with at least one other friend are less likely to get victimized or bullied in the first place. "Friends can be the buffers."

⌐

In any culture you can name, even in nonhuman species, adolescence is marked by experimentation and exploration, by risk taking and impulsivity. And it is marked, indelibly, by interactions with peers. While the process of making friends earlier in life also depends on finding similar interests and attraction, those drivers become far more important beginning in early adolescence. That's why the stereotypical groupings of teenagers into cliques are based in truth. They are repeated so consistently that social scientists have created a taxonomy to describe the social structure of the teenage years. Like the kingdoms, phyla, and classes of Linnaean biological classifications, the labels are organized by size into crowds, groups, pairs, and individuals.[21]

Individuals are the starting point because they are the building blocks. Each brings his or her own personality and interests to any relationship, which may be changeable in the teenage years as identity and sense of self solidify. Then there are interactions between pairs of friends (known as dyads, in scientific terms)—meetings, conversations, and activities over time. Some friendships are short-lived, especially in the shifting social landscape of adolescence, where one's identity begins to diverge from those of the friends you had when you were younger. One boy becomes obsessed with basketball, for instance, while his childhood friend spends all his time playing and composing music.

What so defines friendship in adolescence, however, are the larger alliances. Beyond pairs of friends, there are groups of perhaps three to ten. You could call them cliques. They can be hierarchical, or they can be roughly egalitarian. They can be tightly knit or looser and more porous. A crowd is larger, one of the generalized categories that so dominate the mythology of high school life that *High School Musical* created a scene (in the cafeteria, naturally) depicting the jocks, the brains, the skaters, the theater kids, and so on. While the movie created a Disney version of reality, it wasn't all that far from what most teenagers experience.

What's interesting is the way kids move among these groups—or don't. Juvonen has spent a lot of time examining diversity and the challenges of nurturing friendships across different ethnicities and classes. In schools with notable ethnic diversity, she has found that kids tend to define themselves in multiple ways. "It mixes the reference groups," she says. A similar principle could apply even in places that are less diverse, like Finland, where Juvonen grew up, and some parts of the United States. In those more homogeneous places, diversity comes not from race or class but from having multiple interests. The more multifaceted a teenager's own identity—I play volleyball but I'm also a singer and a scholar—the more likely that child is to have friends in more than one crowd.

⮑

In order to properly understand friendship in adolescence, it is essential to understand that the teenage brain is a work in progress. The last fifteen years have brought a radical rethinking of the particular phases of brain construction that take place between the ages of ten and twenty-five—now considered the span of modern adolescence.

For centuries, it was thought that the brain was fixed at birth. That began to change with the work of David Hubel and Torsten Wiesel in the 1950s and 1960s. Hubel and Wiesel studied vision in kittens, patching and unpatching one eye or the other—and sometimes both—at various developmental time periods that corresponded to infancy, the juvenile period, adolescence, and maturity (those transitions occur in a matter of weeks rather than years in cats). Hubel and Wiesel discovered a critical period early in life when covering up a kitten's eyes could disrupt the development of normal vision, even if there was nothing wrong with the cat's eyes. The problem lay in the brain. Deprived of visual experience, the necessary neural connections failed to form between neurons. The brain was not fixed at all. Instead, it was plastic, changeable by experience.[22] It was such a groundbreaking discovery that it earned Hubel and Wiesel the Nobel Prize.

Once neuroscientists could see inside the brain, they worked out the basic process of brain construction in early childhood. It's a one-two punch, to borrow the phrasing of neuroscientist Jay Giedd. First, there's an exuberant explosion in both the number of brain cells and, more significantly, in the connections formed between those brain cells. The pace is dizzying. Instead of the one thousand new neural connections per second scientists once thought occurred, there are more like one million, according to a 2017 study.[23] That phase of rapid growth is followed by a ruthless process of elimination, a pruning of the connections that didn't pan out, or didn't prove necessary. Brain circuits are very much a use-it-or-lose-it proposition.

This new understanding of very young brains led to a surge of interest in what happens between ages zero and three. The enormous brain growth in those years is critical, and it relies on loving, engaged attention from parents and other caregivers. (Expensive custom-designed brain-stimulating products not required.) Hubel and Wiesel and others concluded that although the brain was clearly not fixed at birth, surely it was mostly done growing soon afterward. By the age of six, the brain is about 90 percent of its adult size.

But it turned out more change was coming. In the late 1990s, Giedd, who coined the one-two punch analogy, led a new kind of project at the National Institute of Mental Health.[24] To that point, imaging studies of children's brains had all been cross-sectional—they captured pictures of brain anatomy at a moment in time. Giedd's new study scanned the brains of a group of children ranging in age from four to twenty. Some kids underwent multiple scans and one was scanned every two years for a decade. The goal was to assess changes in the volume of gray matter and white matter over time. Gray matter is the "thinking" part of the brain, the neurons that do much of the work. White matter, named for the color of the myelin that encases it, is the connective tissue that links neurons together. Giedd and his colleagues discovered something remarkable and previously unknown. In early adolescence, the brain undergoes a second period of growth and change that looks a lot like the earlier spurt in the toddler years. There is a second massive over-production and then elimination of neuronal connections. Maximum neuronal growth occurs at twelve for boys and eleven for girls (the difference is due to the earlier arrival of puberty in girls). Then the pruning process begins again.

This additional wave of development is not quite as substantial as that of early childhood, but it is dynamic and significant. And it makes young adolescents nearly as subject to experience as they were as infants. "Much like Michelangelo's David, you start out with a huge block of granite at the peak of the puberty years, then the art is created by removing pieces of the granite," Giedd said in 2002. "That is the way the brain also sculpts itself."[25]

A second important finding followed from this discovery. The different areas of the brain don't mature at the same rate. Of course, we've long known that teenage brains don't seem fully baked. That's why Robert Sapolsky, in his 2017 book *Behave*, subtitled his chapter on adolescence "Dude, Where's My Frontal Cortex?" The frontal cortex is the part of the brain that is responsible for judgment, planning, and executive function. And it is still under development in the teenage years. That helps to explain why adolescence is such a time of risk taking and vulnerability and why teens are so prone to poor choices—more than 70 percent of teenage deaths are due to car crashes, unintentional injury, homicide, and suicide.[26] It seemed logical, then, to think of maturity as a steady march toward rationality and reasonable behavior. But if that were so, wouldn't younger children, with their even less developed cortices, do even more stupid things than adolescents? Why the jump in risk taking at adolescence that isn't seen before or after?

Because, it turns out, there is something of an imbalance in the developmental force. Adolescence is not just about the slow maturing of the frontal cortex and all the rest of the brain. Instead, it is a period when the brain's limbic system, a critical area for emotion, races ahead and the areas controlling judgment and reason lag behind. What really matters is this gap between thought and emotion.

In 2008 neuroscientist B. J. Casey and her colleagues at Weill Cornell Medical Center first proposed this imbalance model to better explain adolescent behavior and it has since been quite widely accepted.[27] The goal of Casey's work is not to pathologize teenage behavior or complain about how frustrating or frightening kids this age can be. Instead, she wants to understand why they are that way. "I think of it more as an adaptive phase," Casey has said.[28] Experimentation is a natural phase of development, she says, one that is preparing teenagers for the rest of their lives where they will need to be independent of their parents.

The increase in emotional reactivity comes because the limbic regions of the brain—which are responsible for emotion—mature more quickly. They are more susceptible to the surge in sex hormones that washes through the body at puberty. When Casey graphs this out, the picture is clear. She puts age along the x axis and functional development along the y axis. The entirety of the graph is two lines, one to show the maturing of the prefrontal cortex, the other to show the growth in the limbic system. They are not the same. The prefrontal cortex develops in a relatively steady linear fashion. If the graph were a piece of paper, we'd be born at the bottom left corner. Imagine drawing a line from that bottom left corner diagonally up to the top right opposite corner. In reality, brain growth levels off at adulthood—your line will never reach that upper corner. Instead, it will veer to the side of the page somewhere in the middle of development, but the point is that the line initially rises along a gentle consistent slope. The limbic system, on the other hand, undergoes a spurt of development earlier. Its line more closely resembles a shallow bell curve, rising above the prefrontal cortex line and then dropping again and flattening out to rejoin it. The space between the two lines forms a yawning gap. In her papers and presentations, Casey shades that area with hatched lines and adds an arrow pointing to the shaded area, which is labeled "Adolescence."

At the age of fourteen, Ben Steinberg was generally a very level-headed kid. But late one night, he did something foolish. He had spent the evening with a group of friends at another boy's house watching the movie *Happy Gilmore*. Around 2 a.m., it struck the boys as a great idea to sneak out of the house where they were hanging out, run to the nearby home of a girl one of the boys liked, and throw pebbles at her window. But they didn't just wake the girl. They set off the burglar alarm in her house. Then, when a police car showed up, they scattered and ran—a potentially more dangerous offense. When pressed later by his father to account for what he was thinking, Ben said, "that's the problem, I wasn't."

As it happens, Ben's father, Laurence Steinberg, is a Temple University psychologist considered to be one of the foremost experts on adolescent development.[29] At the time of Ben's adventure, Steinberg was overseeing a group of researchers who were thinking about adolescence and juvenile justice. Ben's experience inspired Steinberg to look more closely at the role of friends in the risk-taking behavior for which adolescents are notorious. He suspected that if his son had been alone, he would probably never have sneaked out in the first place and certainly would never have run from the cops.

We know that when they're with their friends, adolescents are more likely to behave recklessly. A teenage driver who has other teenagers in the car is four times more likely to crash than one who is alone. The same is not true of adults. They are safer drivers. Teenagers are more likely to commit crimes when they're together. Adults tend to be alone when they break the law. A teenager's first sip of alcohol or toke of marijuana or experimentation with other drugs is more often in the company of friends than not. Specifically, they are seven times more likely to drink with friends than family and almost never drink for the first time when alone. Most adults think the blame goes to peer pressure— the sometimes overt, sometimes subtle urging by a teenager's friends to try it, to chug, to just have one hit. But Steinberg has shown that it isn't as simple as that. He and his colleagues discovered what they call a "peer effect."

"The presence of peers is so powerful during adolescence that it can make even mice misbehave," Steinberg wrote in his 2014 book *Age of Opportunity*.[30] Pressure doesn't have to come into it, merely presence. When I reached him on the phone, Steinberg explained how they figured it out.

Fittingly, they used a video game. Adolescents and adults came into the laboratory at Temple and brought two friends with them. The game put participants in the driver's seat of a simulation driving game. The goal was to successfully navigate a course as quickly as possible. "Drivers" repeatedly came to yellow lights. Stop or run the light? There were competing incentives. On the one hand, the need for speed encouraged taking risks. On the other hand, drivers were warned, at some intersections, a car will come through just as you enter on yellow. Crashes cost time. So that was a counterincentive to drive carefully, and not push one's luck. To make things more interesting, the researchers promised an extra reward in the form of an additional payment to those who completed the route faster.

Drivers did not make their decisions entirely alone. Sometimes the friends they had brought were in the room with them. Sometimes the friends were in the next room visible on a monitor but unable to communicate with the driver. The results were striking. With friends in the room watching, adolescents regularly took more chances. Adults did not. With friends out of the room but nearby, watching on a monitor but unable to communicate, adolescents still took more chances. In that situation, it wasn't possible for the friends to exert verbal peer pressure, but it didn't matter. "When teenagers knew their friends could see their performance, it increased the amount of risk-taking they engaged in compared to when they were alone," Steinberg says. That did not happen with adults.

Then Steinberg joined forces with Temple neuroscientist Jason Chein and began running the same experiments with the "driver" in a brain scanner. They saw the same peer effect, and now they could also see what was going on in the brain as well. "When kids were in the presence of peers, it activated reward centers in the brain," Steinberg says. "The more that happened, the more risks kids took." The scientists developed a more nuanced theory than one about pressure. "We came to the notion that basically when kids are around other kids it primes their reward system to be more easily aroused and more easily activated. That in turn leads them to pay undue attention to the potential rewards of a risky choice and relatively less to the potential costs."[31]

Well, okay, but what if just knowing the friends are there is still a form of peer pressure? The teenager being tested no doubt suspects that what would impress his friends is to race through the intersections and finish in record time. In anticipation of this, Steinberg, Chein, and their colleagues came up with a way to rule out that possibility. They needed adolescents who wouldn't or couldn't care what their friends thought of them. They used mice.

"We raised mice from the time they were weaned with two other mice from different litters," says Steinberg. "Basically, we created peer groups of mice." Since mice can't play driving video games, Steinberg and Chein gave them alcohol, which triggers reward systems in mouse brains just as it does in human brains. They randomly assigned the mice to be tested alone or in the presence of their peers, and tested half as juveniles (the equivalent of adolescents) and half as adults. How much alcohol would the mice drink if given unfettered access to it? In the presence of other mice, adolescent mice drank more than they did when they were alone. In adults, there was no difference in the amount that they drank.[32] "Unless there's something we don't know about mice that would lead us to think they're thinking about what their peers would want them to do, it seemed to us a reasonable conclusion that there's something about the brain during adolescence in mammals that is hard-wired to be especially sensitive to peer influence and to be more reward-seeking in the presence of peers," Steinberg says. Instead of calling the phenomenon peer pressure, they began calling it "peer presence."

Importantly, peer presence can be a force for good as well as for bad. "When teenagers are with each other, everything that feels good feels even better," Steinberg says. If what feels good is something that also carries some dangers to it, then kids get into trouble because they may pursue that activity ignorant of the dangers—or choosing to ignore them. But Steinberg and his colleagues have also shown that teenagers learn faster when they're with their peers than they do by themselves. And they engage in more exploratory behavior when they're with their peers. Steinberg says, "these could be good things, or they could be bad things depending upon what teenagers are exploring and what they're learning."

Who the peers are becomes very important. "Parents shouldn't worry about peer pressure or peer influence," Steinberg says. "They should worry about who the peers are that their kids are hanging around with." When kids hang around with students who get better grades, their own grades go up over time. Teenagers can also pressure one another not to use drugs. Of course, the reverse is true as well. "Virtually all kids because of the nature of adolescence are going to be susceptible to peer influence and peer pressure," Steinberg says. "The question really is whom are they influenced by and what is it they are being pressured to do?"

It isn't just the brain's reward system that goes into overdrive during adolescence. Puberty also has an effect on the "social brain." Ask a teenager to think about her friends or show him a photograph of an angry person and you'll activate the social brain. Make a teenager feel accepted or rejected—scientists have ways of manipulating this in the lab—and you'll activate the social brain as well. As Steinberg writes, "It's the perfect neurobiological storm, at least if you'd like to make someone painfully self-conscious: Improvements in brain functioning in areas important for figuring out what other people are thinking, the heightened arousal of regions that are sensitive to social acceptance and social rejection, and the greater responsiveness to other people's emotional cues, like facial expressions."[33] In other words, it actually hurts more to be rejected in adolescence than at any other age—not that it's fun no matter how old you are. The bottom line is that teenagers respond to the presence of their peers differently than adults do.

∽

No wonder then that researchers studying a phenomenon known as social buffering found some puzzling results when they studied teenagers. Social buffering is a way of describing the protective, positive effect of one individual on another. It describes the power of one person to reduce another's stress.

The phenomenon isn't limited to people. I mentioned early on that zebra fish showed reduced levels of fear—they froze less—when they could smell the presence of their "friends," and even lower levels if they could see them. This suggested that the phenomenon of social buffering is so important to survival that it has distant evolutionary origins shared by many different species. Furthermore, zebra fish brains might hold clues to how friendship eases stress in humans. In the presence of the shoal, the fish showed patterns of brain activation much like those seen in mammals with friends.[34]

A recent study of wild chimpanzees in Uganda investigated two possibilities. Did the presence of friends mainly help animals weather stressful experiences (like coming across a neighboring group of chimps)? Or did it make a difference to their well-being in a more everyday way, such as when they were grooming each other? The researchers, from the Max Planck Institute for Evolutionary Anthropology, measured stress-related hormones (glucocorticoids) in the chimps' urine in both situations, as well as when the animals were not socializing. Hormone levels were lower whenever friends were together, and the difference was most pronounced in the most stressful situations. Overall, the researchers found evidence that regular time spent with friends led to better routine health.[35] The benefits of social buffering, apparently, aren't limited to moments of acute stress.

Chimpanzees and humans share a neurobiological arrangement called the hypothalamic pituitary adrenal (HPA) axis, which is ground zero for stress responses. Preparing to give a speech to a large crowd is enough to induce stress in most of us, and that stress triggers a cascade of hormones along the HPA axis. The end result is that cortisol levels in the blood rise. Higher cortisol levels get you revved up to perform but can be harmful over time.

When mothers calm their children, it turns out that part of what they are doing is lowering cortisol levels. A group of psychologists at the University of Wisconsin subjected sixty-one girls between the ages of seven and twelve to a battery of stress-inducing tests. (Though the version used for children is adapted to be age appropriate, the Trier Social Stress Test always includes timed public speaking and math problems, an indication of what kind of situation stresses out the majority of us.) After the test, one-third of the girls were reunited with their mothers for fifteen minutes. The moms were allowed to comfort their daughters any way they chose—talking, hugging, and generally being loving and supportive. Another third of the girls got to talk to their mothers on the telephone immediately after they were finished testing. All the comfort came in the form of talk. The third group had no contact with their mothers immediately afterward. Then all of the girls watched a film, after which the researchers measured their levels of stress hormones and oxytocin, which surges upon positive interaction with a loved one. All three groups experienced a rise in the stress hormone cortisol.

Those who interacted with their mothers afterward saw a reduction in those levels—physical touch sped up the process, but voice was enough to have an effect. Those who had no contact with their mothers showed higher levels of cortisol one hour after the stress test. Contact with mothers had an effect on the release of oxytocin as well. It was boosted by mom within fifteen minutes, but there was no change for the girls who did not interact with their mothers.[36]

A team of researchers at UCLA explored this further in 2014. They rounded up twenty-three children between four and ten years old, and thirty adolescents. All of the participants performed two tests with and without their mothers present and then, in a brain scanner, viewed both their mother's face and that of a stranger. The young children put in a stressful situation showed more mature regulation of brain activity when their mothers were with them and when viewing photographs of their mothers.

But how does that response change as kids grow older? That's what neuroscientist Dylan Gee, now at Yale University, wanted to know. She studies how brain circuits mature. Puberty, she found, is a turning point for dealing with stress. Up to the age of ten, mothers calmed down the amygdala by engaging prefrontal circuitry in children's brains that works to control stress. In adolescents, who were eleven to seventeen in this study, Mom's presence no longer worked the same magic.[37] The brain's response to stress remained highly reactive. On the plus side for teenagers, the necessary brain circuitry for managing the stress—a network that connects the amygdala to the prefrontal cortex—is more fully developed, so they are on their way to mature responses. Gee and her colleagues argue that responses to stress are subject to a sensitive period, just as vision, hearing, and language are. They can be molded by experience when children are young. With the right parental nurturing, the positive effects of social buffering can be etched into the brain. But absent that, the brain doesn't learn how to calm itself down.

It seems logical that when parents no longer serve as social buffers, friends might take over, given how important friends are to teenagers. A 2011 study found exactly that in eleven- and twelve-year-olds. The children regularly recorded how they felt about themselves and their experiences throughout their days, and they recorded who was with them. Their cortisol levels were measured as well. Having a best friend present during an experience significantly buffered any negative feelings, lowering cortisol levels and boosting a sense of self-worth.[38]

But things get more complicated later in adolescence. Researchers from the University of Minnesota induced stress in fifteen- and sixteen-year-olds using the same lab test we saw earlier that combines stressors like public speaking and mental arithmetic. Not only did the presence of friends not reduce stress, it made things worse. Cortisol levels went up.[39] At first the scientists were surprised. "We were blown away . . . until we thought about it," says Megan Gunnar, the lead investigator and an expert on social buffering. She realized that the structure of the experiment increased the level of social evaluation because the speech teenagers had to give was about why you'd like to be my friend. "So your friend is actually sitting there helping you evaluate yourself. Oh my God!" Gunnar says with the wisdom of hindsight. "Maybe if you were going in to give a speech about something completely different the friend might have been a buffer. I don't know. We know that they don't necessarily buffer them in the same situations that a parent would."

Gunnar suspects that further investigation of what's going on at this turning point will be very instructive. "Up until puberty, your parents are actually physiologically scaffolding you," Gunnar says. Then that changes. "Parents are supporting you [in adolescence] but they're not in your hypothalamus anymore. They've moved out of your body."

I shared that result with my younger sons, Matthew and Alex, one night over dinner. They were sixteen and fourteen by then, and they were not in the least surprised.

"Of course," Alex said. "Who wants to look silly in front of your friends?"

Matthew added, "It doesn't matter how good a friend it is either."

Back when he was so horrified at the prospect of moving to Hong Kong, Matthew was only just turning nine. Fortunately, I was still able to be a calming influence for him. School started and for the first few afternoons, his sad little face peered out at me through the window as the school bus arrived in our building's driveway. I hugged him each day and told him to give it time. I made sure we had his favorite snacks waiting in the apartment. But then things changed. One day, he clambered off the bus with a boy who lived upstairs. "Can I go to Jason's house?" Matthew asked. Off they went. Soon, Matty had a squad of friends and a routine for the afternoons of getting homework done and meeting up in the playground downstairs before dinner.

Two years later, when it was time to move back to Brooklyn, Matthew had the hardest time of all of us. Just as he had predicted, he hated leaving his new friends almost as much as he had hated leaving his old friends. The second time around, however, he had learned something about himself. He attaches, developing very tight bonds with his friends. Parting is always going to be difficult. But he was older and a little more resilient. Plus, he had home to look forward to.

I had learned something, too. I knew how important Matthew's friends were to him and I made a point of staying in touch with his group from Hong Kong. The following summer, I arranged for Matthew to go to sleepaway camp with four of them, including Jason.

I learned something else as well. Matthew was on to a larger truth. After his first few days in Hong Kong, he was fine. The person who was miserable, I've already confessed, was me. For months. And the reason was that I missed my friends.

FIVE

A Deep Wish for Friendship

t's just after daybreak on a plain at the edge of Amboseli National Park in southern Kenya near the border with Tanzania. In an acacia grove, a troop of yellow baboons is getting an easy start to the morning. A few late risers sleep on in the upper branches, but the others have been dropping down to the ground, one by one. More than half of the nearly seventy animals stroll or sit in the scrubby grass.

"They seem really calm this morning," Susan Alberts says, lowering her binoculars. An evolutionary biologist at Duke University and codirector of the Amboseli Baboon Research Project, Alberts has been observing baboons here for thirty-five years. She keeps up a field biology play-by-play for me.[1]

"There's a nice grooming, a little lip smacking, some embracing, a little inspection."

We watch two females greet each other on a tree branch above us.

"[She] was very relaxed when she was approached. No tension, no anxiety, she didn't cower . . . And now there's a greeting in the opposite direction. Oop!"

For just a second, one female had lunged toward the other.

"The higher-ranking female just reminded the lower-ranking female.

And the lower-ranking female said, oh yeah, I submit. We understand each other."

Our attention turns to a baboon on the ground called Ivy, who is holding an infant. She is approached by Acid, who is pregnant and says a little something as she comes near.

"That's a hello baby grunt," Alberts translates, meaning it is a request to handle Ivy's infant. "Acid is higher ranking than Ivy. You can tell because although Acid was very polite, she wasn't at all shy about it. And Ivy leaned aside slightly. I don't know if you saw it. She just went like that."

Alberts ducks her shoulder ever so slightly then straightens. I hadn't seen it.

Acid moves in front of Ivy, presenting her shoulder.

"A very clear request for grooming."

Ivy and Acid are not related, but, Alberts tells me, after consulting a small binder of field notes, they were born in November 2011 and February 2012, respectively. They are agemates that grew up together.

Now Acid returns the favor and begins grooming Ivy, but she also starts playing with Ivy's baby.

"It's unpredictable, but Ivy and Acid have known each other their whole lives so there's a certain amount of trust there."

You could even say they are friends.

About fifteen years ago, Alberts and her colleagues recognized the strong social bonds among the mothers, grandmothers, and great-grandmothers of these same baboons for what they were—the nonhuman primate version of friendship. More importantly, they uncovered the value of those bonds in the animals' lives, friendship's power to shape outcomes.[2] That discovery was a watershed moment. If humans were the only animal on Earth for whom social relationships affect life span, we would have to search for the roots of friendship in the structure of human society. But since the baboons, along with other species,

share our need to be social, that tells us something different. The wish
for friendship runs deep.

One of the driving ambitions of anthropologists and evolutionary
biologists has been to work out how it was that human beings came
to be the dominant species on Earth, able through adaptation and
scientific ingenuity to live in any habitat, even Antarctica. We know
the process was slow. Seven million years separate humans from the
common ancestor we share with our closest relatives: chimpanzees
and bonobos. There were several critical developments along the way.
When we began to walk upright we could see farther and use our
hands to do more than walk or grip tree limbs. Our ability to make
tools made building possible and changed the odds when hunting.
Language allowed us to communicate as no other species ever has and
therefore to cooperate in wildly impressive ways—ultimately building
skyscrapers and sending rockets into outer space. But another potentially game-changing aspect of the world of early man got very little
attention for a long time: his social life. Trends that leave traces in bone
or carved stones are more tangible. Just because you cannot touch it,
however, doesn't mean the social world didn't touch our ancestors in
formative ways.

In 1976, Cambridge psychologist Nicholas Humphrey wrote a prescient though purely speculative essay posing a new possibility.[3] He
argued that creatures are only as intelligent as their environments
demand. Based on the lives of contemporary Bushmen, early humans
appeared to have a lot of time for sitting around. What would have
challenged their mental capacities? Other individuals. The life of social
animals is like a chess game, Humphrey argued, that required players to be "able to calculate the consequences of their own behavior,
to calculate the likely behavior of others, to calculate the balance of
advantage and loss—and all this in a context where the evidence on
which their calculations are based is ephemeral, ambiguous, and liable
to change, not least as a consequence of their own actions." Social skill,

Humphrey concluded, goes hand in hand with intellect and this need for intellect builds on itself like "an evolutionary ratchet, acting like a self-winding watch to increase the general intellectual standing of the species." His conclusion: "I propose that the chief role of creative intellect is to hold society together."

Humphrey got others thinking. If you leap from intellect to the size of a brain required to contain it, you will notice that monkeys and apes (that is, primates) had much bigger brains relative to body size than most other animals. In 1982 primatologist Frans de Waal published his best-selling book *Chimpanzee Politics*, which noted the Machiavellian scheming required for success in large social groups. The book recounted de Waal's work watching a large population of captive chimpanzees at the Burgers' Zoo in Arnhem, the Netherlands. That research included a pivotal moment that he described in later books. One morning, de Waal witnessed a fight between two chimpanzees. The same afternoon, he spotted the same two animals embracing. It looked very much as if they were kissing and making up. That observation marked the beginning of his interest in the positive side of chimpanzee interactions. Ultimately, he shifted the focus of his studies from aggression to reconciliation, empathy, and morality.

In 1990 two Scottish primatologists, Andrew Whiten and Richard Byrne, took the idea further. They linked primates' capacity for tactical deception and propensity for forming coalitions with the complexity of their societies and the size of their brains. They called their theory the Machiavellian intelligence hypothesis.[4] It emphasized that unlike bees, for example, whose hive has structural complexity with different individuals in different roles, primates live in social systems that involve tight social bonds between pairs of individuals that adjust their behavior according to the subtleties of what's happening around them. In time, since deception was only part of the story, the theory was renamed the social brain hypothesis, which has been popularized by evolutionary psychologist Robin Dunbar of Oxford University.[5]

The social brain hypothesis is based on the idea that nonhuman primate societies have something to tell us about early human societies. It was that suspicion that inspired Louis Leakey to send Jane Goodall to Gombe to watch the chimpanzees. The societies of monkeys and apes in Africa are viewed as indicators of how early hominins might have lived—especially those nonhuman primates that are close relatives or live in habitats similar to those of early humans. Anyone who watches them for any length of time can see pretty quickly that primates have an unusually rich and complex social life. They live in groups with all the drama of those cliques I described from high school.

Living in groups helps animals solve two substantial ecological problems. One is the risk of being attacked by predators. There has always been strength in numbers. A lone individual—whether baboon or human—is a much easier target for a lion than a herd of individuals. The other major problem to solve is finding food. Animals must "make a living" off the land, gathering enough nutritional resources to sustain themselves. There has been a long-running debate within primatology and evolutionary biology as to which problem—predation or foraging—might be primary, and therefore, an animal's first order of business. According to the social brain hypothesis, both problems are solved by first creating a cohesive group whose members can coordinate their actions. The assumption has been that the average size of a group provides a proxy for how hard an animal has to work on its social life, and animals that live in larger groups have been found, on average, to have larger brains.

For a decade or more, the social brain hypothesis was widely accepted. But debate over it never went away and as recently as 2017, a group of primatologists from New York University published a paper using much larger sample sizes and updated statistical techniques to call the idea into question. Their results found that diet, not sociality, predicted brain size—a victory for the camp that has long argued for the centrality of foraging. The authors suggested that perhaps development of social skills followed as brains grew.[6] In an accompanying commentary, British biologist Chris Venditti doubts this will be the last word on the matter. Instead, he believes it is likely to "reinvigorate and refocus" research into cognitive complexity.[7] To know exactly where sociality fits in the evolutionary pecking order, we will have to wait for primatologists to fight it out in the pages of evolutionary biology journals. Already, however, the emergence of the social brain hypothesis has rewritten the story of social behavior in a way that cannot be undone. Sociality is no longer invisible. Whether it is the driving force or a secondary development, it has a starring role in evolutionary theory. "Frankly it's a mistake to search for one and only one cause of brain growth in primates when it seems so clear that there are multiple plausible explanations," says primatologist Robert Seyfarth.[8]

A second important idea, this one from social psychology, was hatched late on a summer night in 1993. Roy Baumeister and Mark Leary sat up talking in the living room of the beach house where they were staying in Nags Head, North Carolina. Everyone else had gone to bed. But Baumeister and Leary were bothered by something.

Even though they were on the beach, they were not on vacation. They were attending an intimate conference of psychologists. There weren't more than twenty-five of them. Mornings were spent listening to one another's presentations and afternoons were devoted to talking inside or out on the beach. The group that week was particularly interested in questions of self and identity. Some of the others were arguing for a new theory that held that humans embrace cultural values that provide life with meaning because of a fundamental fear of dying. (It's known as terror management theory.)

Baumeister and Leary weren't convinced. "Yeah, human behavior is affected by our concerns with dying," says Leary, who is now a professor at Duke. "There's no question about that. But we felt like a lot of the things they were trying to explain with their theory, you didn't need to bring death into the equation. It's much more mundane than that."[9]

If you disagree with a theory in science, it's a good idea to have something else to offer in its place. "We just started talking," Leary says. "If we were going to have one theory in social psychology that was that big, that we thought identified the number one motive that controlled more human behaviors than any other, what would it be?"

They started speculating and, in the wee hours, came up with a new, far more life-affirming idea. An awful lot—if not most—of what people do could be summed up in one phrase: the need to belong. "Human beings," they would go on to write, "have a pervasive drive to form and maintain at least a minimum quantity of lasting, positive, and significant relationships."

At first the idea seemed so obvious that Leary and Baumeister felt sure someone else must have thought of it. No one had, not exactly anyway. Belonging had popped up in psychological literature in various ways over many years, but no one had proposed it as what Leary calls "a master motive."

"It wasn't that we thought the need to belong explained everything on the face of the Earth," Leary says. People need other things. They seek power, achievement, and intimacy. They seek and need material goods. What the "need to belong" theory did was recast behavior that other social psychologists were attributing to concern with death as something that felt more immediate to people's everyday concerns.

Though they were on to something, there were a lot of details to work out. At the time, Leary was a professor at Wake Forest University in North Carolina and Baumeister was at Case Western Reserve in Ohio. Conveniently, though, Baumeister had accepted a visiting position at the University of Virginia for the 1993–1994 school year, and his sister happened to live around the corner from Leary in Winston-Salem. It was natural that Baumeister would make the relatively easy drive from Charlottesville a few times that fall. Every time he visited his sister, he also spent a day sitting in Leary's living room, continuing the conversation they had started in Nags Head.

They covered the walls with giant sheets of paper. Whenever inspiration struck, one or the other would grab a Magic Marker and add to the notes they were creating as they tried to plot out what belonging would have to look like if it could be said to account for a big chunk of human behavior. One side of the room was devoted to what they called criteria. To be a master motive, the need to belong would have to be universal. A lack of it would have to constitute deprivation. It should show up under nearly any circumstance. It should make people feel good (or bad). It should affect how they thought about the world. It should be something people strive for, and it should affect a wide variety of behaviors.

On another wall they assembled evidence in order to bolster their theory with some proof. They cited evolutionary thinking about the need to live in groups in order to defend against predators and improve chances at resources. They included a raft of laboratory studies showing how quickly and easily people formed in-group bonds even when the group they were "in" had been made up by the researchers on trivial grounds (think red shirts versus green shirts). They showed that social attachments form under adverse circumstances, such as military service, where the strength of bonds increases with the intensity of action soldiers see together. They cited proof that people are reluctant to break bonds. And they found early evidence—this was before the explosion in neuroscientific knowledge—of the extent to which cognitive resources were devoted to social relationships.

Finally, on a third wall, they listed implications. What were all the things they could possibly explain about human behavior in light of the need to belong? "We could have written an encyclopedia," Leary once said. They thought their theory explained patterns of group behavior and close relationships, and that group conformity, excuse making, and patterns of self-presentation made sense in the context of enhancing one's chances of inclusion. They mentioned the use of social inclusion as a form of reward and punishment. They saw a role for belongingness in religion and thought they could account for the pursuit of power as arising from the need to belong, too. And they noted "it remains plausible (but unproven) that the need to belong is part of the human biological inheritance." If such a thing could be proven, they said, it would have considerable ramifications for health.

Their paper was published in 1995.[10] Baumeister and Leary hoped their work would make people sit up and say, "We missed this." That is exactly what happened. The theory has been cited (as I write) by nearly eighteen thousand others—an enormous number in the academic realm where one hundred citations is a lot. It wasn't that people had denied the need to belong, but they had certainly failed to appreciate its importance.

Shelley Taylor, a psychologist at UCLA, was aware of the social brain hypothesis and the need-to-belong theory when she sat in a conference audience in 1998 listening to a noted researcher deliver a lecture about the amygdala. Believed to be central to the experience of fear, the amygdala is active when we feel threatened or stressed. Taylor studied stress, so she was curious to hear what this researcher had to say. For decades, the focus of stress research was on the body's fight-or-flight response. The scientist who was speaking studied rats. He mentioned in passing that in his lab they had to house all their rats separately so that the animals wouldn't attack one another. That gave Taylor pause. When she gathered her team later to talk over what they'd heard, someone pointed out that the researcher they'd been listening to, as well as other animal researchers, only studied male rats. It was a rare moment of epiphany.

"The sudden recognition that all of the class theories of stress were based almost entirely on males was a stunning revelation," Taylor wrote later. "I remember thinking, I didn't know there were any big mistakes left in science. We stared at one another as the opportunity that lay before us became clear: a chance to start over and discover what females do in response to stress."[11]

Taylor and her team spent the next several months reading everything they could find about evolutionary stress responses. The result was a new theory, first published in 2000,[12] that went to the very heart of human social interaction and challenged the old way of thinking.

Fight-or-flight was not the only stress response that humans had developed, Taylor argued. There was also "tend and befriend." This was the instinct to nurture and care for others—born of the necessity of raising young—and Taylor argued that it was as hardwired as fight-or-flight, especially in females. Of the centuries of (mostly) male scientists who had written about evolution, Taylor wrote, "In their myopic focus on the aggressive experience of men, they ignore a very rich aspect of both women's and men's lives, namely the caring, nurturant side of human nature. . . . The tending instinct is every bit as tenacious as our more aggressive, selfish side. . . . Tending to others is as natural, as biologically based, as searching for food or sleeping, and its origins lie deep in our social nature."[13]

Hints about those origins were to be found in Africa.

ᘓ

In 1963, when Jeanne Altmann picked up her first pair of binoculars to watch the forebears of the same baboons I saw in Amboseli, she was the twenty-three-year-old wife of biologist Stuart Altmann and something of an accidental primatologist. The Altmanns went to Kenya so that Stuart could observe baboons. Jeanne was primarily planning to take care of their two-year-old son, but her role expanded when Stuart's assistant backed out. The couple spent their days perched on a homemade viewing platform atop a Land Rover while their son played in a jury-rigged playpen in the backseat. It was a challenging year full of car accidents, illnesses, and the Kenyan revolution. But the Altmanns persevered, captivated by the thousands of baboons roaming the acacia woodland and open grasslands along with elephants, zebras, giraffes, and hippos. Over the next decade, they kept returning to Amboseli.

Jeanne wasn't yet a scientist, but she did have a hard-won math degree. She had come of age at a time when women were expected only to marry and raise children. As a high school graduation gift, her grandmother had granted Jeanne's request for a slide rule on the condition that it be small enough to slip into a ladylike purse. Then, as one of three female math majors at UCLA, Altmann had been told that giving the women departmental advisors would be "a waste of time." After marrying Stuart at nineteen and having "a baby before a BA," she finished her math degree in pieces while following her husband, first to MIT and then to the University of Alberta in Canada.[14]

That degree allowed Altmann to combine her quantitative mind and her experience watching baboons to revolutionize the collection of observational field data. "You couldn't even tell how most people did it," she tells me, when we meet in her book-lined corner office in the Department of Ecology and Evolutionary Biology at Princeton University. (I'm a little starstruck. Altmann is now regarded as one of the pre-eminent primatologists of her generation.) The problem with field data, she says, was that the claims being made far outstripped the validity of the ad-lib approach most scientists then employed. Instead, Altmann developed an ordered and methodical technique known as focal sampling that asks the same sets of questions for each animal by following one animal at a time, usually for ten minutes, and recording every-thing that individual does—eating, drinking, sleeping, grooming—and with whom.[15] Her 1974 paper was initially controversial (who was this young woman with no graduate degree?!), but the value of her approach quickly became obvious. Like Baumeister and Leary's paper, Altmann's is within spitting distance of the one hundred most-cited scientific papers of all time. Focal sampling would ultimately allow the valid measurement of social bonds in a way never before possible.

Altmann went on to get her doctorate from the University of Chicago in the Department of Human Development, but the barrier between human and nonhuman research struck her as artificial. "It seemed nat-ural to me that one could get clues in both directions," she says. At the field site in Amboseli, which ultimately became Jeanne's responsibility as Stuart moved on to other projects, she took another pioneering step by focusing on females. "There was this attitude, sometimes explicit, sometimes implicit, that males were where all the action of evolution was," she says. "I felt that particularly in mammals, and even more so in primates, including humans, females had not only control over their own lives—to the extent that anybody does—but also over the next generation. Why should that be irrelevant to evolution?" She also saw that she needed to be in it for the long haul. "It was so obvious that the outcomes came down the road," she says. "The real action was in lifetimes." The only way to capture those lifetimes was to set up a long-term research project that would collect data on the same animals in the same way for generations. The Amboseli Baboon Research Project, begun in 1971, is now second only to Jane Goodall's Gombe for longevity among field sites in Africa.[16]

A second husband and wife team, Robert Seyfarth and Dorothy Cheney, also came to Africa in the 1970s. They, too, were somewhat unlikely primatologists to the extent anyone was "likely" to enter such uncharted territory in those days. Seyfarth was the son of a Chicago businessman, educated at Exeter and Harvard, and expected to go to business school. But he majored in biological anthropology, excited by its mix of genetics, hominid evolution, and primate behavior. Then, in the spirit of youthful adventure as much as anything, Seyfarth went to Cambridge University to study primate behavior with Robert Hinde, who had been such an important influence on John Bowlby. Cheney, too, was a political science major with straight As at Wellesley and a mind like a steel trap. (Her family joked they didn't need bookmarks because they could just tell Dorothy what page they were on.) She put off law school to follow Seyfarth to England. At primatology lectures that year, she was particularly impressed by a speaker who found himself with slides but no audio. He stood in a posh British lecture hall hooting and hollering in imitation of the monkeys he studied. "It was magnificent!" Cheney told me with a grin, and then capably demonstrated by hooting and hollering herself while sitting, incongruously, on the patio of the couple's house in a Philadelphia suburb the year before her death after a long battle with cancer in 2018. She eventually asked Hinde to take her as a student, too.[17]

Their first efforts at observing baboons in South Africa didn't begin well. "After a month of running up and down these mountains trying to follow fleeing baboons, we had some despair," Cheney says. Then Jeanne Altmann sent an early copy of her still unpublished sampling paper. "That saved us. It told us what to do," Seyfarth says. Even though they were getting up at four in the morning, spending hours in the field, and exhausted in the evenings, they began to see that "these were creatures with motivations and strategies," as Seyfarth puts it.

They, too, gravitated toward studying females despite the field's emphasis on males. "It was always the 'phalanx of males, canines flashing,'" Cheney says. "We were more into seeing what the social structure of this society was like. It became clear that all the action was with the females." Practically speaking, there were simply more females to watch. Baboon society is matrilineal, like the macaques on Cayo Santiago. Males come and go, but females stay put. Larger numbers mean more opportunity to consider the variation in behavior among animals.

Variation is of critical importance in evolutionary biology. Darwin's theory of natural selection depends on it. It also depends on the idea that necessity is the mother of invention. "Traits arise or are maintained because they help the individuals who possess them to solve a problem, thereby giving those individuals an advantage over others in survival and reproduction," Cheney and Seyfarth wrote in their book *Baboon Metaphysics*. "A blunt, heavy beak allows a finch to crush hard, dry seed and survive a withering dry season; antlers enable a stag to defeat his rivals and mate with more females. The finch's beak and the stag's antlers did not arise at random; they evolved and spread because of their adaptive value. To understand the evolution of a trait, therefore, we need to understand how it works, and what it allows an individual to do that might otherwise be impossible."[18]

The trait Cheney and Seyfarth zeroed in on was cognition. They wanted to study the evolution of the social mind. Brains and the thinking they make possible are biological traits like any others. If considerable brain power is devoted to thinking about relationships—as it clearly is in humans—evolutionary theory dictates that tendency must come with a benefit. Seyfarth and Cheney set out to identify that benefit in monkeys, for whom social knowledge seemed likely to improve an individual's chances in life. Variation would come in the fact that some animals might be better at it than others.

Evolutionary theories are hard to prove. A common criticism is that they are no better than Rudyard Kipling's "just so" stories, made to fit the facts at hand without actually having to be true. To build a strong case, Seyfarth and Cheney started at the beginning: What did monkeys know about their fellow monkeys? And more to the point, what did they need to know to successfully navigate their social worlds? "If they were to recruit an ally," Cheney says, "how would they know which ally to recruit?"

The couple used playback experiments to approach these questions in a clever, if painstaking, way. Adapted from the study of birds, playbacks either reproduce natural events in a controlled way, or create unnatural situations to see how animals react, and they require patience and ingenuity. Seyfarth and Cheney first tried it with vervet monkeys. They would maneuver a large, heavy loudspeaker (this was the 1970s and 1980s) into an unobtrusive position behind a bush or a tree. At an opportune moment, they played a taped vocalization for a nearby monkey and recorded its response.[19] There was a lot of trial and error and plenty of frustration, but they got it to work eventually.

The aim, to begin with, was simple: Did the monkey seem to recognize the vocalization as coming from a member of its group rather than a rival group? The "three-female experiment" allowed them to explore what monkeys knew about relationships. Could a mother recognize her offspring's cries? To find out, they waited until a baby was out of view, the baby's mother was well positioned to hear a recording, and two other females (the controls) were nearby for comparison. Sure enough, the mothers reacted strongly when they heard what sounded like a fight involving their offspring. Even more interesting: the control females immediately looked at the mothers. "It was as if they were saying, 'that's your kid, what are you going to do about it,'" Seyfarth says. "That suggests that the social relationships were not just a figment of our own human imagination. They were something that existed for the monkeys themselves."

That discovery allowed Seyfarth and Cheney to consider the larger picture of the monkeys' social world that was emerging. "The result of all this social intrigue is a kind of Jane Austen melodrama, in which each individual must predict the behavior of others and form those relationships that return the greatest benefit," they wrote.[20]

In 1986 they put forward a bold idea. Together with Barbara Smuts, who'd been studying baboons in Kenya in the early 1980s, they argued in *Science* that social relationships and social cognition appeared to contribute to evolutionary success.[21] Smuts had already used the term "friendship" in her 1985 book *Sex and Friendship in Baboons*.[22] An account of a yearlong study of relationships between male and female baboons in the Eburru Cliffs troop in Kenya's Rift Valley, Smuts's book went where others feared to tread, at least in its use of the word "friend." Smuts described males and females that spent considerable time together, even though female baboons are only sexually available for relatively brief amounts of time. Two animals named Virgil and Pandora, for example, slept together and groomed together for years. Smuts guessed that such relationships were mainly about reciprocity and not sex, because there just wasn't enough of the latter to keep things interesting. These three primatologists were the first to link friendship and natural selection. But they didn't have much evidence—yet.

In 1992, Seyfarth and Cheney, who were professors at the University of Pennsylvania by then, took over a baboon research camp in Botswana from a retiring primatologist. Radically different from Amboseli, the new camp was in Moremi Game Reserve in the Okavango Delta, which floods annually and becomes a connected chain of swamps. Baboon Camp, as they called it, was on an island in the midst of the Okavango River.

Although it's professionally frowned upon to liken animals to humans, the dirty secret of primatology is that nearly everyone does. "Most of the people I know who are really good at focal sampling and analyzing data and the statistics and everything really enjoy watching their animals," says Cheney. "They can't help but anthropomorphize. They sit around at night and say, Sylvia is such a bitch. Can you believe what she did today?"

That would be Sylvia, otherwise known as the Queen of Mean. And she stood out not only for her meanness. Baboon social hierarchy rivals that of English nobility, and Sylvia ranked right near the top. Think duchess, albeit a malevolent one. She cut a swath through the group, scattering subordinates and biting or whacking animals that failed to move out of her way. But when Sylvia was twenty-one—a ripe old age—her daughter Sierra was killed by a lion. Sylvia became despondent. Sierra wasn't just her daughter, but her closest ally and primary grooming partner.[23]

In her grief, Sylvia did a curious thing. She began approaching other females and grunting at them—an overture that signals conciliation. She then attempted to groom them. A lifetime of irritability isn't easy to overcome; some of Sylvia's intended targets fled in panic when she came near. But still, she worked at it, trying to show her groupmates that she wanted to be friends. Sylvia was behaving a bit like a girl who splits with her boyfriend only to realize her neglected friends have moved on without her.

Maybe, the primatologists thought, there was a way to determine the benefits of these relationships for the baboons themselves—to establish, in other words, what was in it for Sylvia.

Joan Silk knew Sylvia firsthand. Like Ms. Frizzle of *The Magic School Bus*, it's not hard to tell what Silk does for a living. She was wearing a zebra-print dress the first time we met, and a T-shirt covered in monkeys the second time. Her office at Arizona State University is decorated with photographs of baboons, a huge map of Africa, and a line of action-figure-sized primates along the windowsill. But my favorite item was a poster of Charles Darwin reminiscent of the red, white, and blue Obama campaign signs that featured the word HOPE. The Darwin poster, however, says CHANGE and, in smaller text below, OVER TIME.

Studying baboons allows Silk to give in to her natural tendencies. "I'm just basically an incredibly nosy person," she says. "I'm always eavesdropping on the people at the next table. I'm really interested in people's lives. But that's sort of impolite. With the baboons, it's fine. And knowing what happens to them is really addictive."[24]

In the early 1980s, Silk spent a year as a postdoctoral fellow for Stuart Altmann in Amboseli. And she and her family spent a year at Moremi with Seyfarth and Cheney in the early 1990s. Although Silk and Cheney lived across the country from each other, they were not just collaborators but good friends, sometimes e-mailing each other multiple times a day.

In 2002, Silk wrote a paper with a provocative title: "Using the 'F'-Word in Primatology."[25] It was a nod to the fact that "friendship" was still rarely uttered by primatologists professionally even though, privately, they used it freely. Could an animal's social ties possibly warrant the same label we gave to human relationships? Silk asked. Answering that question required rigorous wrestling with the meaning of friendship in humans.

The consensus, Silk wrote, was that human friendships were "intimate, supportive, egalitarian relationships." They required compatibility and an investment of time. Primatologists used the word "friendship" to describe a close and affiliative social relationship, in animals, that required time, support, tolerance, loyalty, security, and equality. What was missing—or not ascertainable, anyway—was the emotional bond, the good feeling that friendship generates in humans. Human and nonhuman relationships are not exactly the same, Silk allowed, but it seemed possible that their similarities could reveal something essential about the attributes of friendship. They would also hint at evolutionary history.

There had already been some debate about evolutionary causes for friendship in humans. Was it a by-product of kin selection, a leftover benefit of sorts? Or are friends a product of reciprocal altruism, favored by natural selection because they provide a basis for trade and alliance formation? The argument for the former held that early humans had lived in small communities mostly interacting with kin and meeting few strangers so that there wasn't much need to distinguish between kin and nonkin. The argument for the latter is logical but doesn't account for the human ability to move beyond tit-for-tat accounting in their closest relationships. The resulting blur of indistinguishable favors that is so common to good friendships was memorably described by novelist Robin Sloan as "a bright haze of loyalty . . . a nebula."[26]

Silk knows just what Sloan means. In preschool, her daughter Ruby started having playdates with a girl named Risa. Risa's father, Brian, regularly took the girls to the movies or for ice cream. "At the beginning, I would drop Ruby off and I would give Brian money for whatever they were going to go out and do. And then the one day, I tried to give Brian money and he said, 'oh, we're past that.'" Silk laughs. "You know how that is, where you stop keeping track?"

Even if monkey relationships could not reach that level of nuance, Silk concluded that they might indeed rise to the level of friendship. If primatologists were going to talk this way, however, they needed more evidence to back it up. They needed to measure the strength of social bonds and the benefits animals reaped because of them.

Fieldwork isn't Silk's forte. "I get lost easily. I trip. I lose concentration." What she is very good at is analyzing large data sets, pulling together reams of information on the individual animals, playing with it, compiling, tabulating, organizing, analyzing. "I call it watching the baboons in the computer," she says. "Trying to figure out what the baboons are trying to tell me about what they think is important."

She called Jeanne Altmann and Susan Alberts in Amboseli.

"Can I use your data?"

By then, the Amboseli Project had complete life histories for more than one hundred females. Alberts had joined the project in 1984 and later become a codirector with Altmann. They agreed to let Silk play with their data.

Choosing the outcome for the analysis was easy. Since reproductive success is the critical evolutionary measure, they would count each female's surviving infants and measure that against what they called a sociality index. It was an assessment Alberts had designed that combined and weighted all social behaviors. "That gave us a number, which reflects the strength of social bonds," says Silk. "It's basically how often females interacted nicely."[27]

After months of organizing the data, Silk came to the moment of truth, when she had to calculate the final result. ("You hold your breath.") She had been quite prepared to see nothing much. Where evolutionary processes are concerned, picking up an effect—a measurable signal in the data—is exceedingly rare. But the signal was loud and clear. Having more and better "friends" was significantly related to reproductive success. Furthermore, strong social bonds mattered more than rank, which everyone had assumed was the most influential variable in the hierarchical world of monkeys. (Remember the macaques at the feeding corrals on Cayo?)

Silk couldn't quite believe it—so much so that she redid the calculation. A lot. "I did it like 50 times. I reanalyzed the data. I tried to correct for all [sorts of] things . . . to make sure that that's really what the data said. It was a pretty stunning result."

Then she called Alberts, who listened, and as a sign of the significance of the findings, asked, "Do you think it's a *Science* paper?!"

It was. Logically, if natural selection has favored the ability to form friendships, then three things must be true: friendships must increase reproductive success, individuals must seek out friends (or try to), and they must have the kind of social intelligence necessary to select the best partners. The results from Silk, Alberts, and Altmann appeared in *Science* in 2003. In the same issue, a paper from Seyfarth and Cheney's team showed that baboons recognize relations within and between matrilineal family members. Taken together, those papers, plus a later report mentioning Sylvia, provided persuasive evidence on all three evolutionary counts.[28]

The primatologists didn't stop there. What exactly was a good relationship, they wondered? The next step was to think about quantifying
the dimensions of social bonds. Silk, Alberts, and Altmann began to
measure the stability of bonds, distinguishing between relationships
that depended on events in the moment and those that held steady
over time. They also considered the overall tenor of social bonds—
were they mostly positive (grooming and proximity), mostly negative
(aggression), or a little of both? Females with the strongest, highest
rates of affiliative interactions turned out to enjoy the most balanced
grooming relationships (literal versions of "you scratch my back and I'll
scratch yours") and the most stable relationships. Strength and stability were strongly correlated—if you found one, you were likely to find
the other. And the strong, stable relationships were the ones that led
to good things—more surviving babies, in this case. In a nutshell, the
primatologists found that the quality of a relationship mattered most.[29]

When they read the 2003 Amboseli paper on the strength of social
bonds, Seyfarth and Cheney called Silk to suggest replicating the
results with data from Moremi. Silk never said no to more data to play
with, but she also wanted reassurance. "I used to wake up in the middle
of the night and worry about the Amboseli result," she admits. She was
sure that it was correct for Amboseli, but "it didn't actually have to be
true about the world."

But the result held. Again, the strength of females' social bonds was
the most important factor in their reproductive success.[30] (In Moremi,
they measured not the number of surviving infants but the length of
the babies' lives.) "It gave me a lot of confidence that we had found
something that was really meaningful about these animals," says Silk.
Alberts agrees. "When they replicated it, we really knew it was true."

Then Silk, Seyfarth, and Cheney went a step further. In 2010 they
reported that females that had strong stable social bonds didn't just have
more babies, they lived longer themselves.[31] A few years later, a study led
by Elizabeth Archie, one of Alberts' students and now one of the associate directors of the project at Amboseli, found the same thing.[32]

Friendship has since been found not only in baboons and chimpanzees and other primates but further along the mammalian evolutionary chain in elephants, hyenas, whales, and dolphins.[33] In some of these species, it is the males that stay put and the females that migrate. In such cases, the males tend to have the stronger social bonds. Other species, such as chimpanzees and dolphins, live in fission-fusion societies, in which individuals spend long periods of time alone or in small groups (the fission), coming back together with others to mate (fusion). Even here, scientists see evidence of strong ties between specific individuals.

"There's growing evidence that what natural selection is favoring is the formation of strong social bonds," Seyfarth says. He and his colleagues have arrived at a relatively simple view of what kind of bond is required. It must be strong, stable, and relatively equitable. "That's what friendship is: a long-term, positive relationship that involves cooperation." That may not be as poetic as Aristotle's discourses on the perfection of *philia*. But this straightforward approach has allowed scientists to set up a base camp from which to launch further explorations into the still uncharted territory of relationships.

One intriguing implication of this work, for instance, is that it blurs the long-standing distinction between friends and family. This is ironic because baboons spend the great majority of their time with relatives. But the fact that they can and do build useful friendships with nonrelatives is the exception that proves the rule—the strength and stability of a bond can be more significant than its origin. As loneliness researcher John Cacioppo once noted, "the relationship with your spouse can be positive and supportive, or it can be the most toxic that you have in your life."[34] By this logic, relatives and sexual mates can be considered friends, but only if the bond is rewarding. Kinship gives you a head start, in a sense, on the necessary investment of time and the probability of compatibility, but family togetherness is no guarantee of joy.

At the end of our morning in the field with the baboons, Susan Alberts and I perch on a small rise. Zebras graze on the plain in front of us and there's a wildebeest off to the left. In the far distance, a plume of dust appears. It's a piki-piki, a motorcycle that the Kenyans use as a bush taxi and the first sign of people since we arrived at dawn.

That first paper on social bonds changed everything, Alberts says. It made plain that studying baboons had explicit parallels to humans. She is convinced the stripped-down version of relationships in animals have a lot to tell us about human interaction. "That's where you learn the most about friendship," Alberts insists. "That's where you begin to understand it. All human friendships are characterized at base that simply."

There once was a baboon named Kathryn, she tells me, that lived to the impressive age of twenty-six. Because she had no offspring, Kathryn had no relatives left by the time she was about sixteen. "She formed friendships with other single females," Alberts says. "There's no way they weren't friends. They spent time together. They groomed each other. They were in each other's proximity. And that's because potentially the servicing of mutual social needs leaves you both better off. And that's the point of friendship."

Alberts reminds me of the two female baboons we watched earlier. "You saw Ivy and Acid," she says. "They're not related. They're relaxed around each other. They've known each other forever. They have a pre-dictable relationship that gives each of them comfort. Yeah, sometimes there's tension. . . ."

She pauses and looks me in the eye for emphasis. "If that's not friendship, what is?"

SIX

The Circles of Friendship

My first year of college, I became fast friends with Sara, one of my roommates. For Christmas the year we met, I wanted to buy her a teddy bear (my nickname for Sara, for reasons I can't remember, was Ted). While I was shopping for gifts with my father, I mentioned this plan. He immediately led me to the toy department and we hugged all the teddy bears until we found one we agreed was suitably soft and wonderful. The gusto with which Dad entered into the task was a little out of character—he was loving, but not sentimental, and I didn't think he had ever hugged any of my teddy bears. Still, I was thrilled to share the choosing of a meaningful gift. Sara loved the bear and she loved that my father helped pick it out. That teddy bear lived in our various dorm rooms throughout college.

Fifteen years later, when I was thirty-two years old, my parents went on a bicycling trip in Nova Scotia, Canada. While pedaling along a quiet road, my father had a heart attack and died. The shock of the loss was compounded by its unexpectedness. Dad was sixty-seven. There had been no warning signs. When I saw him for what turned out to be the last time a month earlier, at my son Jake's first birthday party, my father had seemed fit and hearty.

My grief felt enormous and disorienting—like crossing into an altered world. A gaping hole had opened in front of me, yet the people of the world were oblivious. They didn't see the hole and just kept walking.

The memorial service was held in my hometown of Philadelphia where my father was a respected attorney and civic leader. Many, many people attended. Family, friends, colleagues, elected officials— even the mayor—were there. But what meant the most to me was how many of *my* friends traveled to Philadelphia. My oldest friend, Arianne, who had known me since I was a baby, came all the way from London. Those who couldn't attend called or wrote. They could see the hole. They came and stood at the edge with me, acknowledging its depth and the seemingly unbridgeable distance to the other side.

Among those who came was Sara. She brought back the teddy bear.

"I think you should have this now," she said.

⌐

At its best, friendship makes you feel valued and supported. It stretches out a net when you need to be caught. In that span of fifteen years, Sara and I checked the box on each of the fundamental characteristics of the relationship that anthropologists find appear most regularly across the cultures of the world. Friendship makes you feel good. It includes a willingness to help, especially in times of need. And it often involves the giving of a gift, to signal the value of a relationship (though the gifts, like cut flowers, need not be of lasting value).[1]

Even considering these elements, there is no one way to do friendship. That makes sense for a behavior guided by natural selection: multiple strategies can lead to success. Sociability depends on appetite and opportunity. Friendship varies according to age, stage, and to some extent, gender. Yet there are limits to the possibilities, and persistent themes recur. I have come to think of the landscape of adult friendship as variegated, like foliage that can come in every shape and size from a pine needle to a palm frond, but must, because of the chemistry of photosynthesis, always include shades of green.

The word "friend" is not categorical like "cousin" or "colleague." *Friend* carries emotional weight, signifying something about the quality and character of a relationship that is based on history and the content of repeated interactions, just as Robert Hinde recognized. That repetition is important. A pleasant interaction with a stranger at the supermarket doesn't make that person a friend. For most of us, friendships are voluntary, personal, positive, and persistent, and they usually assume some measure of equality. Although friendship can encompass betrayal, jealousy, and other negative emotions, if we call a spouse or relative a friend, we do so to signal the quality of that relationship, its extra special character.

Some people are most bonded to people they have known for years. Our friend Dan, he of the allergic reaction in chapter 3, grew up with a tight group of buddies. Decades later, those men—some with families, others single, some still living in New York City, others not—are still the people closest to him. The comforting base of their shared childhoods was deepened in their twenties by tragedy. Two of the group, brothers, lost their father on September 11, 2001. The ensuing days, weeks, and years of being there for one another worked like chemical fixation to set the group's bond permanently. A friend I'll call Catherine has a very different set of bonds. She has never married and has lived and worked in Hong Kong, London, and New York. She describes her "forever" friends as a set of "small lights around the world" who don't all know one another, but who have proven consistent in their determination to stay in her life, and to make her feel like a valued part of their own lives. In both of these examples, the context of the relationships varies, but the content is the same.

There is chemistry in friendship as there is in romantic relationships. We find it rewarding to learn that others like us and agree with us, and we tend to respond by liking those people back. But a calculus—sometimes unconscious—is at work as well. A friendship's rewards should outweigh its costs; the satisfaction and commitment we derive should be greater than the investment we make and the alternatives we forgo. A concern about fairness hovers in the background. The best relationships make us feel good and provide us with reliable sources of assistance and succor. Friendships that don't do this tend to fall away.

They fall away for other reasons, too. Shifts in friendship come with age and biology—sociologists call these the "turnings" of life. The twenties are what writer Ethan Watters dubbed the "tribe years," at least for urban college-educated professionals.[2] New people are easier to meet, old friends (like college roommates) might well live nearby, and there is time for hanging out, for drinks, dinners, softball leagues, and spa days. The thirties, on the other hand, are sometimes described as the decade where friendship goes to die, killed off by marriage, children, jobs, relocating. Mismatched friendships—one has kids, the other doesn't—can be especially hard to sustain. When Jake was eighteen months old, my husband and I took him with us to the wedding of another of my college roommates. After being home with a baby, I was looking forward to a weekend in the company of old friends. Instead, Mark and I spent hours keeping Jake occupied while waiting for our childless friends to wake up, finish their leisurely brunches, and commit to an activity we could conceivably join. By the forties, most everyone has work responsibilities, and/or houses full of school-age children who need to be fed, transported, and generally tended. As a result, we still don't prioritize friends. Now, as I pick my head up from raising a family, I must say that the fifties and beyond look promising.

Intriguingly, a study using seven months of cell phone data tracked relationships across the life span.[3] In this limited context, midlife friendship fared well. Out of 3 billion calls, the researchers identified 2.5 million male callers and 1.8 million females and tracked who they called most often. They labeled "best friends" as the most frequent conversational partner of the same sex and nearly the same age. (Yes, those could be homosexual romances or twin siblings, but mostly they won't be.) Call length and frequency were assumed to indicate emotional closeness, and the researchers kept track of who initiated the call. They found a gradual shift in young adulthood from calling parents to calling friends. Romance surges—calls to opposite sex same-age peers take a growing fraction of time through the age of 28. You can almost *see* people ditching their friends for love. But then at 30, friends get the great majority of phone time—though that might be a sign of less opportunity to meet face-to-face. Between 29 and 45, there is decreasing communication by cell phone with a "spouse," presumably because you now live in the same house, while more time is spent talking with best friends, especially among women. Toward the end of life, there is more balance in calls among three generations, which presumably includes both same-age friends, children, and grandchildren.

We can glean more about friendship by combining big data sets like this with personality results. Psychologists scale personality along five dimensions: openness, extroversion, agreeableness, neuroticism, and conscientiousness. A college student who loves to travel and meet new people is probably high in openness and extroversion. A writer who meets every deadline is high in conscientiousness and perhaps neuroticism, too. Yet where one falls on these personality spectrums doesn't tell you as much about friendship as you might expect. There is little evidence, for instance, that similarity in personality has the same attraction as shared worldviews do.[4] When more than twelve thousand Britons provided information on up to three close friends, most lived within five miles of each other, saw each other daily or weekly, and were not more than two years apart in age. (Just over one-quarter of the friends listed were biological relatives.) The exception to these rules were for people who were high in openness to new experiences. They were most likely to have friends who lived farther away, were substantially older or younger, or whom they didn't see regularly but still considered close. Agreeableness and to a lesser extent, extroversion, were associated with more traditional friendships that thrive on stability and proximity.[5]

But there are other ways that we bring our individuality to bear on our friendships. While conducting oral histories with sixty-three senior citizens in the 1980s, sociologist Sarah Matthews of Cleveland State University recognized three distinct styles of friendship: independent, discerning, and acquisitive. Independent people consider themselves self-sufficient and content to socialize casually. Their friendships are often circumstantial, formed with schoolmates, coworkers, or neighbors, but not maintained. "I'm my own man. Do I have friends now? I have people that I know," said one such man. Discerning people are deeply tied to a few very close friends. Those relationships tended to be long lasting and harder to establish later in life. One man counted only two men he had met in his twenties as true friends because they had "an impact" on his life. That's a pretty high bar. Acquisitive people, by contrast, collect a variety of friends as they move through life. They are open to meeting new people, but keep up old relationships, too. "Unless you make friends, you're isolated," one such woman told Matthews. "You have to make a conscious effort."[6]

A team of German psychologists recently used Matthews's friendship styles to study nearly two thousand adults over forty.[7] The categories held up well except that acquisitive friendships seemed to more properly divide into two groups: the selectively acquisitive, who were choosier, and the unconditionally acquisitive, who were mainly interested in socializing broadly. This larger study also showed that the discerning friendship style was most common, and the independent style the least. Education, physical health, years living in the same place, and the number of daily contacts with friends all predicted friendship style. More education, for example, often leads to better social skills, and higher incomes make it easier to socialize more widely (at restaurants and concerts, say). On the other hand, poor physical health can cause people to retreat emotionally or suffer reduced mobility.

Most of what I've described so far, however, is based on what social scientists call WEIRD societies, those that are Western, educated, industrialized, rich, and democratic. Their citizens represent 12 percent of the world's population but as much as 80 percent of study participants.[8] Anthropologist Daniel Hruschka of Arizona State University set out to gather what he could find about friendship more broadly from hundreds of societies. He had his work cut out for him. Friendship has been ignored in anthropology as in every other discipline. Where it is mentioned in the ethnographic record, it is usually as an aside. Yet Hruschka found something like friendship everywhere he looked.

Trobriand Islanders in the South Pacific used a codified system of friendship to trade safely throughout their ring of islands. Travelers carried bracelets and rings symbolizing specific friendships. The jewelry and the relationships were passed down from father to son for generations. Lepcha farmers in Eastern Nepal say that one of their gods thought up friendship when he was drunk, as a category of relationship with people who had what he did not: Indians for copper vessels, Tibetans for rugs, Bhutanese for fine cloth. But the Lepcha also describe friends as those who look after one another in emergencies, help with farmwork, promise to care for one another's children, and offer travelers a place to stay. Their friendships are both economic and affectionate. Women on the island of Crete say that their female friends provide a haven for sharing problems, lessening worries, making mundane tasks more fun, and helping one another manage in a male-dominated culture. Western Apaches, when they first met white people, found the foreigners' friendships to be "like air," insubstantial because they could form and dissolve relatively quickly. The only society Hruschka found that didn't have friends was the *atevi*, who have no word for trust or friendship, but ten words for betrayal. The atevi, however, exist only in science fiction, in the *Foreigner* series by C. J. Cherryh.[9]

A thought problem known as the Passenger's Dilemma has provided a window into cultural variation in friendship. It goes like this:

You are a passenger in a car driven by a close friend, and he hits a pedestrian. You know that your friend was going at least thirty-five miles per hour in a zone marked twenty. There are no witnesses. Your friend's lawyer says that if you testify under oath that your friend's speed was only twenty miles per hour then you would save your friend from any serious consequences. What would you do?

Dutch social scientists put that question to thirty thousand white-collar workers from more than thirty different countries. The answers varied dramatically by nation. Fewer than one in ten Americans would lie to protect a friend when breaking a law was involved. Most northern Europeans said the same. Among the French and Japanese, the number willing to lie rose to three out of ten. Venezuelans were most likely to choose loyalty to friends over society's rules—seven in ten would lie.[10] Those willing to lie in the Passenger's Dilemma were more likely to lie in other social situations as well. A doctor might lie about a friend's health to reduce the friend's insurance premiums. Or a food critic might embellish a review of a friend's restaurant.

Only one of the several theories proposed to explain these cultural differences stands up to scrutiny and it concerns economic and political uncertainty. "Material help among friends becomes more important in societies where daily life is more uncertain," Hruschka says. For each of the countries the Dutch researchers included, Hruschka collected data on the level of confidence in the rule of law, the presence or absence of corruption, and perceptions of stable government. He found a strong correlation between uncertainty and the probability that people will lie for friends. Countries like the United States and Switzerland are low in uncertainty and low in friendly liars. More uncertainty, such as in Cold War Russia, brings more lying (more than half of Russians from that era chose friends over civic institutions). The results are thought provoking but leave unanswered questions. Is it uncertainty that makes us value friends more? Or is a willingness to break the law to help friends the cause of broader social uncertainty?[11]

⌒

There's a reason we speak of our "closest" friends. As we will see in chapter 9, the phrase reflects how our brains work and the blurring of self and other that occurs when we think of our loved ones. Not surprisingly then, metaphors of spatial proximity turned up as another theme in Hruschka's work. In Bangladesh, they describe "thick" friends. In Mongolia, they speak of "inside friends."[12] Social network scientists use more prosaic terms—they speak of strong ties or a "core discussion network" of those we rely on to discuss important matters.

It is common to envision our social relationships as a series of concentric circles. This is another idea popularized by British anthropologist Robin Dunbar but the model was created by psychologist Toni Antonucci, who works down the hall from Jim House at the University of Michigan. Back in 1980, as the first large epidemiological studies were showing a connection between health and a crude count of social relationships, Antonucci and her colleague Robert Kahn wanted to measure the quality and complexity of those relationships as well as their quantity. They conceived of a "social convoy," a protective layer of friends and family traveling with us through life. "Each person can be thought of as moving through the life cycle surrounded by a set of other people to whom he or she is related by the giving of social support," they wrote.[13] And they noted that they were extending John Bowlby's notion of attachment to "the world of the adult"—just as Bowlby himself had done.

When she began to test her idea, Antonucci and her team presented hundreds of people with a piece of paper showing three concentric circles with the word "you" at the center. They instructed participants to place family and friends within the appropriate circle (close, closer, or closest). The very closest are those you cannot imagine life without.

The next group is slightly less central, but important nonetheless. Even relationships in the outer circle include the giving and receiving of affection, aid, and affirmation.[14]

Very few people qualify for our innermost circles. The exact numbers have been counted in multiple ways by a variety of researchers—social network scientists, repeated national surveys, and so on. The average American claims about four close social contacts, and the great majority of us have between two and six intimates. In one survey, only 5 percent of Americans listed as many as eight close contacts. At the other end of the spectrum, another 5 percent said they have not one close relative or friend.[15] Again, education tends to result in larger networks: college graduates have nearly twice as many people in their inner circles as those who dropped out of high school. For everyone, the number of close confidants usually decreases as we age, which isn't as alarming as it sounds. (More on that later.)[16]

I must pause here to debunk a myth. In 2006 a report in *American Sociological Review* triggered fears of a "loneliness epidemic" and garnered dramatic headlines like "Friendless in America" and "The Lonely American Just Got Lonelier."[17] Three sociologists had compared the results from the 1985 and 2004 versions of the General Social Survey. This biannual survey of Americans consists of personal interviews with about fifteen hundred people and has been repeated roughly every other year since 1972. The 2006 drama centered on this question: "Looking back over the last six months, who are the people with whom you discussed matters important to you?" The researchers found that between 1985 and 2004, the percentage of people who had discussed important matters with *no one* tripled, from about 8 to about 25 percent. Among those who said they had spoken with someone, the average number of confidants dropped from about three to about two.

Except . . . the 2006 paper wasn't quite right. There were methodological differences in the 1985 and 2004 studies that created the appearance of a major jump that wasn't really so major. In some cases, the answers reflected not having any "important matters" to discuss rather than not having any person with whom to discuss them.[18] A Pew Research Center reanalysis of the data found that the direction of the trend was correctly reported but that the number discussing important matters with no one moved only from 8 to 12 percent, not 25 percent.[19] Another study also turned up 12 percent as the number of those who claim no close confidants.[20] Counted differently, when asked a more direct question, the percentage of Americans with very few close friends has been remarkably consistent for decades—almost always in the single digits.[21] This does not mean that loneliness is not a problem for those who suffer it. It should be clear by now just how dangerous loneliness is for your health. But the repeated reports that cite this particular study (read carefully and you find it everywhere) are not accurate.

If 90–95 percent of us do have close confidants, then who are they? Out of four intimates, two or three are likely to be relatives—50–75 percent of the inner circle, in other words.[22] Like the baboons of Amboseli and Moremi, we have a head start on a strong bond with family members since we spend so much time with them. But more than the origin of a relationship, its quality best predicts health outcomes. The people we rely on most should be those who make us feel valued and supported. An early study of a group of Russian Jewish immigrants living in Venice, California, found that though they had left behind or lost what family they had, they survived by developing close ties with nonrelatives, "in essence synthetic families."[23]

Unless the marriage is miserable, a spouse or significant other, if we have one, is usually part of the inner circle. "In adulthood spouses are the most important adult relationship," says Bert Uchino, the expert in ambivalent relationships we met in chapter 3. "You spend a lot of time with them. Your sense of self becomes more embedded with who they are as well." This is especially true for men, who tend to focus more of their emotional life on their wives and let other friendships fall away. Women are more likely to maintain or develop other close friendships in addition to a spouse.[24]

But both men and women today often call their spouses their best friends. We do it because the phrase "best friend" telegraphs something about the strength of the relationship that *wife* or *husband* does not. That habit, however, is relatively new and far from universal. In the 1980s researchers in Jacksonville, Florida, asked participants whether they referred to their spouse as their best friend; 60 percent said they did. They asked the same question of people in Mexico City. Almost no one said they called their spouse their best friend.[25] This does not mean that the inhabitants of Mexico City necessarily had any less affection or love in their marital relationships. They simply regarded the relationship as distinct from friendship. It is increasingly the case, however, that in Western society, we expect a spouse to be a soul mate. Psychologist Eli Finkel has called this the era of the "all-or-nothing marriage."[26] The difficulty in finding such a person no doubt helps to explain the rising numbers of people who are single. In a 1965 survey, three out of four college women said that so long as a man met all their other criteria, they would marry someone they didn't love.[27] Today, concerns about job security and shared views on child raising still topped a Pew survey of what the never married are seeking, but very few people say they would marry without love.[28]

Family members do tend to have staying power in that close inner circle. Ninety percent of us have at least one sibling of some sort and they are likely to be the only people with whom we have a lifelong relationship.[29] The moment my husband and I found out we were having a third boy, Mark's response was to celebrate the bond these three brothers might have throughout their lives. But there is a reason for the popularity of the adage: "You can choose your friends, but you can't choose your family." Being related is no guarantee of affinity. Some relatives grate on us, or pick fights with us, or simply don't share our interests. Often we can live with such differences, but not always. The last five years have seen an increase in studies of family estrangement. It is more common and more complicated than you might think. One study found that 8 percent of some two thousand Britons had cut off a family member and 19 percent said they or another relative were no longer in contact with family.[30] In the United States, 7 percent of adult children reported being detached from their mothers and 27 percent from their fathers.[31] Divorce, abuse, addiction, or other variations of bad behavior can all be driving factors. Even when family relations are happy, we lose loved ones to death or disease. Both of my parents were once in my inner circle, but my father is dead and, as I write, my mother has severe Alzheimer's disease. One of the most difficult aspects of my mother's condition was realizing that I had lost her as a confidant even though we still spent a lot of time together. Loss can be countered with gain, however. I have become much closer to one of my aunts as we navigate my mother's decline together.

In truth, the makeup of our social convoys shifts and changes over time more than we may realize. But there are signs that we have consistent social "signatures" that reveal our patterns of intimacy over time. Finnish computational scientist Jari Sarämaki teased those out using cell phone data. He first enlisted 24 British students ages 17–19 who were on the verge of moving from high school to university. Three times over 18 months, he had them list the names and phone numbers of relatives as well as friends and acquaintances from school or work, and then rank everyone according to emotional closeness. Then the researchers tracked the students' calls. On average, the closest contact (very often a parent) got one-quarter of a person's calls and the top three contacts got 40–50 percent of all calls. Those calls lasted longer, too. The rest of each network saw high levels of turnover. Yet each person stayed true to his or her basic tendencies across calls and texts—they communicated a little or a lot. They called mainly Mom or their three best friends from high school. Or they called all their new college friends more than their old friends. New college students might be expected to be in a time of transition, but when the researchers expanded beyond them, first to 93 people and then to a database of over 500,000, the same findings held true. [32]

Most of us do not have only four friends. Those in our outer circles—and beyond—are an important part of our lives, too. Scientists approach the study of those people a little like they do the physics of a spider web. How are these networks built? What forces can they withstand? The web analogy breaks down, though, when it comes to purpose. Spiders build webs to catch prey. People build networks to thrive and connect.

Two essential ideas underlie social network analysis: connection and contagion. Who is connected to whom? Each individual is considered a node in a larger network. Remember the party game Six Degrees of Kevin Bacon in which you try to connect yourself to the actor within six steps? One of my sons was in preschool with a boy whose father was related to Kyra Sedgwick, who is married to . . . Kevin Bacon. Got it in five. That's a network of sorts, though it won't provide me any sustenance. I could also map my former journalism colleagues at their current jobs or the families whose children play basketball in Brooklyn. Contagion refers to what, if anything, flows across the network ties: job tips, germs, dollar bills, or the names of excellent basketball coaches. Connection and contagion equal the structure and function of a social network.

In 1938, what appears to be the very first map of social networks charted the friendships in one Vermont village (population: one thousand). It was created by social scientists from Bennington College mainly to show it was possible to depict what they called "the circulatory system" of a community, a phrase that nicely sums up how vital relationships are.[33] In the published paper, relationships were drawn by hand, with individuals as circles, and arrows indicating friendships. Nearly every family in town (94 percent) participated. Socioeconomic status, occupation, family size, church and club membership, and favored reading material were all recorded, but what the researchers most wanted to know was who people said they liked. While there were limitations, they thought they created "a highly reliable account of the friendship nuclei . . . of this village."

Members of the village cliques had similar levels of wealth and status—though there was a tendency to aim high in naming friends. The entire village fell, with some overlap, into "seven well-defined constellations plus a ring of isolated and semi-isolated individuals." Each constellation centered on its most popular person. One chart, dubbed "the hub of the village," consisted of well-off business and professional families. The matriarch at its center, the village's "lady bountiful," received an impressive seventeen nominations as "best friend." (She named only two herself.) In other groups, merchants and their wives hung out together. Factory workers, many of them Catholic, formed a cohesive group around the wife of one of the skilled workmen. A woman who ran a boarding house was named as a friend by the wives of two farmers, a chauffeur, two salesmen, and a bank teller. Of the small group of unconnected residents, who named no one and were named by no one, most had only recently moved to the village and reported active social lives in other places. But three were noted as potential "social problems."

From such humble beginnings grew the work of Nicholas Christakis and James Fowler. Christakis began his career as a physician and became interested in the way one person's illness took a toll on another, especially a spouse. That led to the realization that pairs of people connect to other pairs, as he puts it, "to form huge webs of ties stretching far into the distance."[34] Captivated, he became a sociologist specializing in social network analysis and joined forces with Fowler, a political scientist. It was an unorthodox career move, but in conversation, Christakis, who now runs a social network lab at Yale, crackles with ideas and energy and seems entirely capable of shooting off in unlikely directions, intellectually and professionally.

Christakis and Fowler argue that our social choices are like building blocks that shape our networks: "Do you want one partner for a game of checkers or many partners for a game of hide-and-seek? Do you want to stay in touch with your crazy uncle? Do you want to get married, or would you rather play the field?" they write.[35] But our networks also shape us—your position in your family, whether your friends are friends, and how many friends your friends have can change the course of your life. "Each and every one of these ties offers opportunities to influence and be influenced," they write.[36] One of the main takeaways of the book that Christakis and Fowler published in 2009, *Connected*, was that they were able to trace the influence of friends three degrees away (halfway to Kevin Bacon) on everything from voting behavior to obesity.

Christakis and Fowler did much of their work with a previously unappreciated gold mine of data on friendship: the Framingham Heart Study. When Christakis went looking for large data sets, he discovered that Framingham participants had filled out sheets listing close friends and family as emergency contacts, but no one had ever made use of them. In total, there was a record of more than fifty thousand ties.

To see how an emotion traveled through a network, one of the contagions they studied in Framingham was happiness. The happiest people were those with the most social connections, even if some were not intimate friends. It seems that if you smile, indeed, the world smiles with you. Daily exposure to small moments of happiness—your neighbor shouting a jolly hello, the barista remembering your name and your order, beginning your workday with a chat with a colleague about your favorite television show—can combine to raise our mood. Christakis and Fowler found that happiness is slightly more contagious than unhappiness. By their calculations, each additional happy friend boosted your cheeriness by 9 percent, while each additional unhappy friend dragged you down by only 7 percent. It follows then that the more links you have in your network—the larger your wider social net is cast—the happier you will be. "If you're at the center of a network, you are going to be more susceptible to anything that spreads through it. And if happiness is spreading more reliably, then on average you're going to be catching happy waves more often than you catch sad waves," Fowler told the *New York Times*.[37]

Happiness isn't all you get from those you don't know well. In 1973, Mark Granovetter wrote one of the most influential papers in sociology on "the strength of weak ties."[38] He argued that while strong ties

provide support, weaker ties—to coworkers, friends of friends, neighbors we nod at—provide a bridge. They connect us to information and
ideas we would miss if we interacted only with our intimates. Granovetter asked a group of adults who had recently found jobs through contacts how often they saw the contact around the time they got the tip:
55.6 percent said occasionally, 27.8 percent rarely, and less than 20 percent often. Paradoxically, tighter, more cohesive groups may ultimately
prove less resilient to change because of the lack of outside contact if
they don't maintain weak ties. In other words, a little new blood can be
healthy. Granovetter's work has led to the widely held view that weak
ties are good for new information but not support, and vice versa for
strong ties.

Reading about exchanges of information, I think of my recent e-
mail exchanges with my friend Kat. We are fellow science writers and
meet up every year or two at conferences where we share drinks and
maybe dinner. We enjoy catching up, but months can go by with no
contact between us. Yet when she needed advice on hiring help for
her elderly parents, she e-mailed me because she knew that I had
done it for my mother. The friends who are most useful in such situations are usually those who have been there. A few years earlier, when
elder care was all new to me, I had relied in turn on my friends Jenny
and Julie, both of whom had mothers with Alzheimer's disease, but
also my friend Stephanie's friend Vera, the quintessential friend of a
friend, whom I have met exactly once, but who has a lot of experience
in this realm.

Had I thought to tell him about this chain of conversations when
we spoke, I imagine Harvard sociologist Mario Luis Small would have
nodded his head in recognition. He thinks there is more to weak ties
than providing information. He finds we routinely confide in people
we don't know well, even as we also maintain a core network. "People's
true pool of confidants is everyone they run into," Small writes.[39] He
conducted an in-depth case study of thirty-eight graduate students
over the first year of their programs—it is full of detailed concerns and

conversations about finances, career prospects, and family life. This small-scale intimate approach reveals truths about people's lives that a big data approach will miss, says Small. His strategy is to ask people who they *think* they talk to, and then to get at what they actually *did*. But he supported his findings with a national survey of more than two thousand adults eighteen and over. In both the large and small studies, more than half the time, "[people] often confided even deeply personal things in people they weren't that close to."[40]

One reason we do this is to explicitly avoid our usual intimates. "The guy who has cancer doesn't want to tell his wife because he doesn't want to worry her. Or you don't want to tell your mother you're broke because she's broke too, and she'll try to send you a check," Small says. Second, people look for others with similar experience or professional expertise. That could be a doctor or a therapist, or a relative stranger. "People favored empathy more than they feared being hurt—like the guy who's sitting at the lobby at the day care waiting for his kids and sees the other guy has a tan line on his ring finger and is also getting divorced. Next thing you know, they're venting," Small says. The third reason is the simplest of all. "They just talked to the person because they were there."

We don't recognize how often we do this. "We're less self-protective than we think we are," Small says. He has come to believe that while sharing matters, with whom we share matters less than we thought. "Mere expression—literally just talking—makes a difference. In practice, there are going to be very few things that we're willing to tell friends that we won't have told at least one other person." Some situations—like dinner conversation or long plane rides—make listening more likely, he points out. "We respond to context at least as much as we think rationally about who we ought to be turning to."

Small is not trying to overthrow the orthodoxy regarding the importance of strong ties. "If you're in trouble and you fall, the safety net, which is your small circle of confidants, is going to catch you," he says. But he believes a steady stream of social interaction is as important.

"The people who are really in trouble are not the people who can't name their three or four safety net people, they're the people who are literally not running into anybody on a regular basis."

⌒

My friend Elizabeth once told me about spending an evening with her then-boyfriend and his brother. I will call them Richard and William. They were watching a football game. The brothers, who I think would describe themselves as best friends, sat close to each other on the couch, but it didn't seem to Elizabeth that they said much more than "pass the popcorn." Yet when Richard and Elizabeth went home, he said, with great satisfaction, "It's so great catching up with William."

She and I had a good laugh when she told me this story. That is not what catching up looks like to us. Our friendship is lived differently. We see each other and a third friend, Amy, mainly in the summer, during a few weeks of vacation on a small island off the coast of Long Island. Our three families flow in and out of one another's homes and on and off the beach throughout the days and nights, and Elizabeth, Amy, and I talk. We talk while we walk for miles on the beach or sit on the sand in the soft light of late afternoon. We talk on my deck or on Amy's screened-in porch. We talk over coffee or wine. We talk about our eight children, our husbands and ex-husbands, and aging parents. We talk about work, what to make for dinner, and what to read next. By the time we go home, we are thoroughly versed in the specifics of one another's lives. Catching up, in other words, requires conversation, and lots of it.

The relationships I have just described are almost comical for how cleanly they fit the stereotypes of how women and men do friendship— face-to-face versus side by side. Historically, though, if you consult Aristotle and Montaigne, it was men who believed they were most capable of deep friendship. "Men have friends, women have acquaintances," went a quote collected in Calcutta by a social scientist in the 1960s.[41] Oh, how times have changed. Recent decades have brought a strong view that women excel at friendship and men are duds. In a conversation about female friendship on the TED stage, Jane Fonda and Lily Tomlin celebrated their close bond, which began when they starred together with Dolly Parton in the movie *9 to 5* in 1980. "Women's friendships are like a renewable source of power," Fonda said, and then added, "I feel sorry for men."[42]

I agree with Fonda about the (potential) glory of women's friendships, but the story of gender and friendship is not quite so simple as women good and men bad. For one thing, women can be awfully nasty (*Mean Girls*, anyone?), and the men in my life have very strong friendships. My husband, Mark, and Amy's husband, Tom, are every bit as close as we women are. They do spend a lot of their time together engaged in sports, running, or playing tennis, but they talk while they compete. "At the end of three sets of tennis with Tom, I feel completely at peace," Mark told me. "It's a respite. There's no judgment."

Researchers have pulled together the hundreds of studies examining friendships among men and among women. What are the expectations? How do people interact and behave? When the results from fifty quantitative studies on self-disclosure were compiled, women proved only a little bit more likely to share.[43] Jeff Hall, a professor of communication studies at the University of Kansas, has looked extensively at gender differences in friendship. "Men's and women's [styles of] intimacy are different, but men's and women's satisfaction are not that different, and the value of friendship is not that different," he says. "Men and women define the importance of friendship in a very similar fashion. They want to have friends who are authentic and loyal and trustworthy equally."[44]

It occurred to Hall that we build our friendships based on our expectations. "We cultivate the complexity of our friendships," he says. When he combined multiple studies and analyzed the resulting larger sample of 8,800 men and women, he found many similarities but a few differences. Men are more likely to build relationships that focus on what friends can do for them, what opportunities they can open, what kind of resources they can provide—all of which Hall describes as agency.

Women, on the other hand, are more likely to expect their closest friends to offer emotional nourishment and support.[45]

Some of those differences are cultural. As soon as psychologists began studying friendship seriously in the 1970s and 1980s, they realized that girls were being raised to seek out one-on-one friendships, while boys were learning to socialize in groups, such as Boy Scouts or Little League. Some of those cultural traditions have changed: today, boys have more playdates and girls participate on more sports teams. But many of us—male and female—turn to different people to fulfill different needs, a phenomenon that social psychologists Elaine Cheung and Wendi Gardner of Northwestern University dub *emotionships*. "When we're in a particular emotional state, we have specific people on tap for that," Gardner explains. "I would call my sister if I was sad. I would never call my sister if I was anxious because she and I are too similar."[46]

Two strangers sit in a laboratory with an assigned task: get closer to each other. To help them along, they refer to a set of thirty-six questions. The first few are easy. *Given the choice of anyone in the world, who would you want as a dinner guest? Would you like to be famous? In what way?*

The next couple of questions start to dig a little deeper. *What would constitute a "perfect" day for you? When did you last sing to yourself? To someone else?*

As the exercise continues over forty-five minutes, the questions become increasingly intimate. *How do you feel about your relationship with your mother?*

The final questions work more explicitly to create closeness between the two people in the lab. *Tell your partner something that you like about him or her already. Share a personal problem and ask your partner's advice on how he or she might handle it.*

The questions were developed in the 1990s by social psychologist Arthur Aron and his colleagues at the State University of New York at Stony Brook.[47] Aron studies romance and needed a way to facilitate closeness in the lab. His questions worked remarkably well—one couple got married. The exercise is now known as the "36 Questions That Lead to Love."[48]

But the two people working through the questions in the video I'm describing aren't looking to fall in love. They are two straight men, already friends, participating in a study of emotional closeness in men. The men were in the laboratory of Beverley Fehr, a psychologist at the University of Winnipeg in Canada, who studies friendship. She wanted to see whether creating a situation where opening up was expected would change the stereotypical response. "I was wondering whether men would be comfortable engaging in this non-normative behavior or whether it would be like taking bad medicine. You really don't like doing it, but it has a beneficial effect in the end."[49]

Even brought along gently, the men in the study hesitate. As the pairs of men take turns asking and answering questions, there comes a predictable moment. "One person will be reading the question," Fehr says, "and the other guy will be sitting there. He will very often look completely stunned. There's usually a pause and a very common response at this point is to use the F-word or to say 'that's deep.'" The men look like "deer in the headlights," yet go on to reveal very personal information. One admitted to wondering whether he had been adopted.

In addition to the lab sessions with questions, Fehr had men play sports together and watch sports together—both rituals of male bonding. She asked the men to describe their levels of closeness before and after all three conditions: watching sports, doing sports, and disclosing personal details. Preliminary results indicate that men felt closest after the thirty-six questions exercise—they rated their levels of satisfaction with the friendships higher. Those who did sports together also reported higher levels of closeness. Simply watching a game together did not change how men felt about their friendships.

Male or female, discerning or independent, we are each faced with a major constraint when it comes to friendship: time. There will never be more than 24 hours in a day or 168 in a week. In adulthood, myriad competing priorities mean we spend less time with friends. When 300 individuals—from teenagers, middle-age adults, and retirees—were asked where they were and what they were doing every 2 hours, teenagers were with friends 30 percent of the time while 40- to 65-year-olds logged just 4 percent of their time with friends and the retirees 8 percent. Every age group associated the presence of friends with pleasure. Married people were happier when they were with their friends than with their spouses, although to be fair, that could reflect that some time spent with spouses is spent doing chores. What made people happiest was to be with both spouse and friends. The researchers concluded: "With friends our attention becomes focused, distractions lessen, awareness of time disappears: We emerge into a world in which the intimacy and joy shared with others is the fundamental reality, and for a time the world becomes a different place."[50]

When time is limited, our relationships are, too, no matter how many people we know. You simply cannot be in contact with all the people in your life to the same degree every day. Jeff Hall wondered something basic: How much time does it take to make a friend? To find out, he surveyed 355 adults who had relocated within the last six months. He asked each person to identify a potential friend they had met when they moved. Where did you meet? he asked. How much time did you spend together last week? In a typical week? What kind of friendship or acquaintanceship would you say you have with this person? What do you do together? Do you hang out, work, talk? In a second study, Hall caught freshmen and transfer students before they had a chance to make friends. Three weeks after arrival, he asked them to name two new people they had met other than roommates. Roughly three weeks and six weeks later, he checked on the new relationships, asking students to add up the estimated hours spent with each person and report whether and when the relationship had gotten closer.

Hall was looking for "cut points," estimates of the amount of time necessary to bring a new person closer. Combining the two studies, Hall found it took between 40 and 60 hours to move from an acquaintance to a casual friendship, from 80 to 100 hours to call someone a friend, and over 200 hours of togetherness before someone rated as a best friend.[51]

How people spent those hours mattered. By itself, time is not enough, nor is proximity. "I had people in the first study who spent in excess of 400 or 600 hours together with somebody at work and still only called them an acquaintance," Hall says. "We don't like everybody we work with and we don't want to be friends with everybody that we work with." Hanging out and eating together were good for turning acquaintances into friends. The ways that people talked to each other mattered. "When you spend time joking around, having meaningful conversations, catching up with one another, all of these types of communication episodes contribute to speedier friendship development," Hall says. "Think about what it does if you and I are casual friends and the next time I see you, I say what's been going on with your life? You catch me up. That action is meaningful because it says that whatever is happening in your life, I want to bring into the present in my relationship with you. Consider how many people you don't bother to ask. You wander into the office, you say hey, and that's that."

While self-disclosure is often viewed as critical in relationships, Hall found it wasn't the only thing that mattered. "It doesn't have to be intimate," he says. "When we focus too much on [that,] we're neglecting the value of joking around with one another and seeing what's going on with each other. It's not that self-disclosure doesn't matter. It is that other things do, too." Even playing video games appeared to bring college students closer together, as did watching television and movies. "Who are we to judge that if we have friends over for a game night, we are so much superior than our teenagers playing video games together?" Hall tells me. I confess to having been very judgmental about video games. Hall laughs. "I think it's about both/and. It's not that either you spend time talking and joking around and catching up or you only play video games. Both are friendship-developing activities."

Fifty hours struck me as a high bar. Surely there were people I had bonded with more quickly? Then I remembered Aristotle's observation that friendship takes time even if the wish for friendship comes quickly. What does fifty hours look like in real life? A college student could come more than halfway to that requirement by taking a class with someone for three hours per week. "Accumulating 30 hours is not hard if you are a college freshman," Hall says. "It's super hard if your life is like mine when I have a wife and kids and a job, and my closest friends are hard to come by and time is of the essence." *I'm right there with you*, I think.

Hall has a theory that the conversations that bond us to others require emotional energy. "They also take time and they come with risk," he says. We are willing to take that risk, Hall believes, in order to satiate our need to belong, the very idea put forward by Baumeister and Leary. Once that evolutionary need is met, we begin to conserve energy—to talk less, engage less. Ultimately, we privilege the relationships that offer the most bang for our energetic buck—those that make us feel the greatest sense of belonging—and we engage in the type of talk that gets us there. It isn't enough to want friendships. "You have to spend time investing in people," Hall says. "It's important to keep it in mind as a priority. It's clear that many, many adults don't feel they have a lot of time, but if we do not prioritize these relationships they are not going to develop."

SEVEN

Digital Friendship

On New Year's Eve 2010, artist Tanja Hollander was sitting at home in Auburn, Maine, communicating with some friends. To one, who was working on a film in Jakarta, Indonesia, she sent instant messages. To another, who was stationed in Afghanistan, she was writing a letter with paper and pencil. The juxtaposition got her thinking about her relationships with these two people from very different worlds, both of whom had significant roles in her life.

That led to thinking about the friends and family she had recently photographed for a project and then she started scrolling through Facebook. She noticed that her 626 Facebook friends lived all over the country and even all over the world. These 626 people represented many different parts of her life—relatives, childhood friends, college friends, fellow artists. Some she had known for decades, some for less than a year. Some were close to her; others were almost strangers.

The seeds of an idea planted on New Year's Eve germinated a few days later when another pair of friends came over for dinner. "We were talking and having little fights about politics or art," Hollander remembered later. "I realized that what friendship was to me was either going to somebody's house or having your friends over for dinner and getting into these arguments and drinking too much red wine but still being friends in the morning."

She decided to launch a new project. It would be called "Are You Really My Friend?" The goal was to visit each and every Facebook friend in person and, if they were willing, to make portraits of them in their homes. It took Hollander six years. She traveled more than 200,000 miles visiting 34 states and 12 countries. She contacted all 626 Facebook friends and made a spreadsheet recording how she knew each person and for how long she had known them. Sixty-seven did not respond, 53 didn't schedule to meet, 21 said no, 14 were businesses, 13 either unfriended her or she them, 5 canceled, and 2 were no-shows. But everyone else said yes. Some were remarkably generous, inviting Hollander to stay for days, cooking her meals, showing her around their hometowns. She visited 424 homes and produced 430 portraits, as well as thousands of photographs documenting her travels on social media. She wrote a book and made a movie.[1]

The resulting exhibit, which I stumbled upon during a visit to Mass MOCA, a contemporary art museum in northwestern Massachusetts, examines the boundaries between our public and private lives, our online and off-line worlds. It is a remarkable exploration of the meaning of friendship in the era of social media. While Hollander found that she wasn't necessarily friends in the fullest sense with everyone at the beginning, entering their homes and visiting with them in person—exactly what we don't do on Facebook— brought her closer to each one of them. Yet those deeper connections would never have happened if she hadn't been connected to these people via social media to start.

Are you really my friend? It is an existential question for our time. Has the overuse of the word "friend" devalued the relationships it describes? In a connected world, how valuable are our online relationships? Are we, in fact, disconnected from one another? Must we visit someone at home to call that person a true friend? As a society, we are pondering the effects—psychological and physiological—of the digital age on relationships and on our psychological health. We worry about screen time for children, about video games for teenagers, about the distracting presence of cell phones, and about whether texting is making it harder to string a sentence together or communicate face-to-face.

The technological change brought by mobile phones, the internet, and social networking sites feels as if it has hit us with the speed and force of a tsunami. Cell phones began proliferating in the 1990s. By 2018, 95 percent of American adults were using them. Smartphones, which added instant access to the internet, were introduced in 2007, and now more than three-quarters of us have them. Eighty-nine percent of us use the internet. There is near saturation for all things digital among adults younger than fifty and among higher-income households. Nonusers tend to be older than sixty-five, poor, or residents of areas with limited service (six in ten rural Americans say access to broadband internet is a problem where they live). Between 2005, when the Pew Research Center began tracking social media use, and 2018, the proportion of Americans using social media to connect, to keep up with the news, to share information, and for entertainment, went from 5 to 69 percent—that means it jumped from one in twenty adults to seven in ten.

Of all social networking sites, Facebook has been the behemoth—and has dominated the science of social media as well. Roughly two-thirds of American adults were on Facebook in 2018. Teenagers once were, too, but when their parents and grandparents began showing up in their feeds, young people shifted to Instagram and Snapchat and may have moved elsewhere when you read this. But their general online presence has only grown. In 2018 fully 95 percent of teenagers used smartphones and close to half were online "almost constantly."[2]

Because social media is so new, the science investigating its effects is also new—perhaps about where research into the effects of loneliness and social isolation was twenty-five years ago. However, this research begins with a different set of assumptions. We were surprised to discover that social relationships could affect our health, whereas it seems we would be surprised if social media didn't. Fear and panic over the effects of new technology dates back to Socrates, who bemoaned the then-new tradition of writing things down for fear it would diminish the power of memory. Thomas Hobbes and Thomas Jefferson both warned that communal relationships would suffer as industrial societies moved from rural to urban life. "Before we hated Smartphones, we hated cities," write sociologists Keith Hampton and Barry Wellman, who study the effects of technological innovation.[3] Radio, video games, and even comic books have all caused consternation. Television was going to bring about the dumbing down of civilization. When people learned I was writing a book about friendship, they regularly asked what I would say about the effects of social media.

Given what we know so far, what I have to say is this: friendship, real friendship, hasn't changed much. It is alive and well, even thriving in some senses. That said, just as we have to be sure to make time for friendship in our busy lives, we would do well to remember how important face-to-face interaction is in cultivating and maintaining relationships. Proximity matters.

One reason for optimism about friendship is that many hysterical headlines about digital technology like "Have Smartphones Destroyed a Generation?" and the (often only slightly) more sober scientific reports that give rise to those headlines are not about relationships.[4] They grapple broadly with the effect of technology and social media on "well-being." The definition of well-being varies, but it can encompass depression and anxiety, loneliness, fear of missing out (FOMO), measures of happiness, and, yes, relationships, including but not limited to friendship. It is beyond the scope of this book to cover all of those possible effects in depth. Furthermore, the results to date have been so mixed they amount to a scientific version of he said, she said. For every study that finds a rise in loneliness, there is another showing an increase in connection. Yet as this book was going to press, the situation was becoming a little clearer. The results are mixed because the effects are mixed, and not nearly as dramatic as advertised. The newest research should calm the hysteria around social media (and help us zero in on the more nuanced problems that do exist). It also allows us to begin teasing out the ways the digital revolution has specifically affected friendship.

First, let's consider the big picture. When you step back, your perspective on the research changes as if you were viewing a pointillist painting up close and then from across the room. A few years ago, psychologist Jeff Hancock, an affable Canadian who runs the Social Media Lab at Stanford, was perplexed by the way one day's news story on the effects of social media often conflicted with the next report, even when both cited work from his lab. Hancock decided to look at everything that had been done. In a major meta-analysis, he combined data on the effects of social media on all the possible measures of well-being from 226 papers published across 12 years beginning with the first such report in 2006. That provided a sample size of more than 275,000 people. The result, which he presented at a conference in 2019, is clear. "Using social media is essentially a tradeoff," Hancock says. "You get very small but significant advantages for your well-being that come with very small but statistically significant costs." What's more, the overall effect on well-being, meaning the amount of variation among individuals that could be attributed to technology use when all effects were combined, was "essentially zero," Hancock says. To be specific, it was 0.01 on a scale in which 0.2 is considered a small effect size.[5]

That result is echoed in other studies. Andrew Przybylski and Amy Orben, experimental psychologists at the University of Oxford, published a trilogy of papers early in 2019. In the most striking of their studies, they rigorously analyzed data on more than 350,000 adolescents and showed persuasively that, at a population level, technology use has a nearly negligible effect on adolescent well-being, explaining less than half a percent of the variation. For context, they compared that result with some other information gleaned from the data. Eating potatoes is associated with nearly the same negative effect as technology use, and wearing glasses is worse for teenage well-being. Yet no one is wringing their hands over potatoes or eyeglasses.

These new studies reveal the pitfalls of the statistical methods scientists in this field have employed to date. Some studies have been sloppy. Nearly all are correlational and must rely on participants to report on their own use, which is notoriously unreliable. Most studies are also unintentionally biased by what the researchers expect to find. That is because most analyses are done with large existing data sets like Monitoring the Future, an ongoing study based at the University of Michigan that tracks the behavior, attitudes, and values of teenagers. Those data sets are full of all kinds of bits and pieces of information (that's what allowed Orben and Przybylski to make the comparisons they did). Using such data makes it easy to turn up tiny, statistically significant results that don't actually mean very much and often conflict with other results. To drive home this point, Przybylski and Orben calculated that had they followed standard statistical operating procedure, they could have produced roughly 10,000 papers showing negative screen effects, 5,000 indicating no effect, and another 4,000 demonstrating positive technology effects on young people—all from the same group of data sets. "We're trying to move from this mindset of cherry picking one result to a more holistic picture," Przybylski says. "A key part of that is being able to put these extremely miniscule effects of screens on young people in a real-world context."[6]

Importantly, it is not that social media is never a problem. But the problems that exist are more nuanced than the science has shown. Effects really do depend on the user—age and mental health status are just two important factors that make a difference. "It's not one finding or recommendation fits all," says Ariel Shensa, the statistician for the Center for Research on Media, Technology and Health at the University of Pittsburgh. Someone with a tendency toward depression or anxiety is at more risk from social media use than someone who begins with a higher level of psychological well-being. But very few studies focus on the people with underlying problems. "We know that problematic smartphone use may as likely be a result of mental health problems as a cause and that calls for a different set of solutions," says clinical neuropsychologist Tracy Dennis-Tiwary of Hunter College in New York. "That means we have to focus our research efforts and our resources towards discovering those solutions."[7]

A science of social media 2.0 will require more precise questions. The overwhelming attention to time on social media (both frequency and duration of use) ignores content or context. The very concept of screen time is essentially meaningless given the variety of possible ways to occupy that time. Who and what we interact with matter as much or more than for how long.

But let's get back to friendship. If you separate the research on social media's effect on friendships and relationships from other measures of well-being, the news is mostly good. Metaphorically speaking, the rich get richer. Those with strong social networks off-line add to them online. Digital communication is one more channel of connection. It is, however, also possible that the poor—socially speaking—may be getting poorer in that existing loneliness is aggravated by social media use. Even here, though, the data is mixed and clearly depends on age.

To dig a little deeper, let's look more closely at Hancock's meta-analysis. He found that research into the effects on well-being generally fall into six categories: depression, anxiety, loneliness, eudemonic happiness (finding meaning in life), hedonic happiness (enjoyment in the moment), and relationships. If you separate the results for each effect, there are significant but small negative effects for anxiety and depression, though not loneliness. At the same time, there are significant but small positive effects for life satisfaction and relationships. Out of all six categories, the effect on relationships is the largest and it's positive (0.19, just under the 0.2 that counts as a notable, if small, effect). When it comes to relationships, benefits win out over costs. This is especially true for older adults for whom additional online connections are strongly positive. "For older people, social media is a real avenue of connection, of relational well-being," says Hancock. Those older adults who use social media report more support from both their grown children and their nonkin friends.[8]

The first—and still only—major survey examining the intersection of people's social media use and their relationships both online and off-line was conducted for the Pew Research Center in 2011 and led by Keith Hampton, who is now at Michigan State University.[9] Despite the worries over degraded and devalued relationships, the Pew survey of 2,255 adults found that people who were more active on social media had stronger relationships across the board. For instance, Facebook users had slightly more close relationships than nonusers. They got more social support—receiving advice, companionship, help when sick, and so on—than nonusers. "Someone who uses Facebook multiple times per day gets about half the boost in total support that someone receives from being married or living with a partner," the researchers wrote. The Facebook users were also more politically engaged and more trusting.

Notably, Facebook served to revive "dormant" relationships. (A recent article in the *Atlantic* put a different spin on this. It was titled "Facebook: Where Friendships Go to Never Quite Die.") Whether you think this is a good trend or an annoying one, the persistence of these old relationships is new. It is one of the few places where Hampton sees a true historical difference in the effects of today's technology versus previous innovations.

For centuries, changes in technology have equaled changes in mobility, which in turn can lead to a reduction in the density of relationships. Before the Industrial Revolution, small-town life in agrarian America meant that people were born and died in the same community. They had very close, dense networks of relationships. Those were useful for some things, like social support and building a barn, and less good when it came to pressure to conform. Mobility, which first came with the move to cities, made it easier to find people with shared interests. It reduced ostracism. It allowed for flexibility in networks.

Ties became less permanent as people moved through the life course, perhaps changing cities after graduation or moving from job to job. Mobile phones and social media did not begin this change and people are not now suddenly giving up place-based relationships for new types of online relationships.

But there are a few significant changes from the past and friendships that never die is one of them. This stems in part from a continuity that is now routine. You keep the same mobile phone number for decades and perhaps the same e-mail address, and your presence on social media stays put even when you don't. "We no longer predictably lose social ties from high school or when we leave one job and go on to another job or move from one neighborhood to another," says Hampton. As a result, access to social support, stress, and opinion formation are all likely to be affected. (For the record, the other significant change that Hampton has identified is our increased awareness of what everyone around us is doing, which amounts to an informal social surveillance that harkens back to the gossip of small towns.)

The persistence of relationships is evident in Hampton's 2011 Pew study where the highest percentage of Facebook connections were from high school. The average Facebook user had 229 friends and the breakdown looked like this:

22 percent from high school
12 percent extended family
10 percent coworkers
9 percent college friends
8 percent immediate family
7 percent people from voluntary groups
2 percent neighbors

Another 31 percent couldn't be classified into any of these categories. When the researchers dissected these numbers, they found that a very small number might be considered strangers or people who had only met in person once. While these long lists of Facebook friends obviously extend to the outer limits of people's social networks and beyond, 40 percent of users had friended all of their closest confidants. In other words, they were friends online and off-line with those in their closest circle (I will return to this idea).

The 2011 Pew study also served as an update to the question of social isolation and the number and diversity of Americans' close social ties. Pew's more limited 2008 effort, which I mentioned in chapter 6, helped to correct the idea that as many as one-quarter of American adults had "no one with whom to discuss important matters." The more accurate figure then was not 24 percent but 12 percent. But by 2011, that was down to 9 percent. A similar pattern held for the number of close confidants the average American claimed. It went from 1.93 in 2008 to 2.16 in 2011. Neither study provided any evidence that mobile phones or internet use was negatively affecting Americans' close relationships. In most cases, use of the internet and cell phones was associated with larger and more diverse social networks. Social media users were less likely to report having no close confidants than nonusers (5 percent versus 15) and reported more close social ties (2.27).

Positive effects of social media are clearest when looking at social networks. The average American had a total of 634 ties—a figure that includes our closest friends and family, as well as many far weaker ties. Technology users had substantially bigger networks with an average of 669 social ties, while nonusers had an average of 506 ties. Furthermore, the more someone used the internet, the larger that person's network tended to be.

The Pew researchers had to conclude that social media use seemed to support intimacy, not undermine it. "There is little validity to concerns that people who use [social networking sites] experience smaller social networks, less closeness or are exposed to less diversity," they wrote. In fact, "a deficit of overall social ties, social support, trust and community engagement is much more likely to result from traditional factors, such as lower educational attainment."[10]

The flip side of friendship, as we've already seen, is loneliness. I was surprised that Jeff Hancock's analysis turned up absolutely no effect either way for loneliness. A large 2017 study from the University of Pittsburgh found something different. They surveyed a nationally representative group of nearly 2,000 people about duration and frequency of use on eleven different social media platforms as well as well-being. Those who spent the most time on social media were twice as likely to feel lonely, and those who visited social media sites most frequently had more than three times the odds of having higher perceived social isolation.[11] But this study does not tell us which comes first: loneliness or social media use. This is the same chicken-or-egg problem that Jim House faced with his 1988 paper on social relationships and health that I mentioned earlier in this chapter. It is possible that individuals who are already feeling socially isolated tend to subsequently use more social media, perhaps as a form of self-medicating. "Those with fewer 'in-person' social outlets may turn to online networks as a substitute," said senior author Brian Primack when the study was published. On the other hand, it could be that more social media use causes people to feel increasingly isolated. It is important to establish which way it works.

I hope I have convinced you by this point in the book that friendship does indeed run deeper than Facebook. At the heart of the definition of friendship that biologists and sociologists have established is the acknowledgment that we treat our friends differently than we do acquaintances, and we differentiate between close friends and less intimate bonds. If we take our off-line concentric circles of connection and overlay them onto our online networks, it is usually the case that we are in touch with our closest friends and family in multiple ways. "Most people's online relationships are, in fact, relationships that are formed around real existing physical places that then migrate online," Hampton says. "They're school relationships, kin relationships, workplace relationships. The vast majority of online relationships start offline." Those that do originate online and become close also migrate to the real world. "We're talking about a network of relationships that are maintained through multiple media. We talk to them on the phone, we see them in person. We share emails with them. And we are friends on Facebook," Hampton says. The more media we use to maintain a relationship, the stronger that bond is likely to be.[12]

The global label "friend" doesn't mean that people don't know the difference between close friends and distant acquaintances. We are smarter than that. Another group of researchers at Michigan State found that use of Facebook supplemented off-line relationships by providing another channel for relationship maintenance. Users proved perfectly able to distinguish among the kinds of relationships within their circles, reporting that 25–30 percent of Facebook friends were "actual friends."[13] Lists of friends grow because people hesitate to delete old connections, but the closest friends stand out from the crowd.

Now for a note of caution. Digital technology may expand our networks, but it also disrupts more private moments. I know quite a few people who are unable to put their phones down, absorbed in their social media feeds when they should be absorbed in their companions, or unable to carry on a conversation without punctuating it with Google searches. This is not just a problem for teenagers; adults are equal opportunity offenders.

We are not imagining the disconnect that the constant presence of cell phones can create. In a 2012 experiment, researchers wanted to know whether the mere presence of a mobile phone, sitting off to the side, would affect the quality of a conversation between two strangers. Asked to discuss "an interesting event that occurred to you over the past month," the participants reported achieving less closeness when a phone was present (versus a notepad in the control condition), especially in a conversation that was meaningful rather than casual.

Phones interfere with the connection between parents and very young children, a fact that should remind us of Bowlby's work on attachment, where we started this story, and that has implications for other relationships. Tracy Dennis-Tiwary and her colleagues re-created a classic psychology experiment called the still-face paradigm. Designed to evaluate the effect of reduced interaction between babies and depressed mothers, the experiment has mothers and babies play together, then the mother stops responding to the child, holding her face still. This part is distressing for the babies and difficult for the mothers. Psychologists are especially interested in what happens during the final reunion period when mothers reengage. How quickly do the children recover? Dennis-Tiwary's team substituted a cell phone for the still-face phase of the experiment. They had parents spend two minutes staring at their phones. Losing their parent to a digital world proved just as distressing for most of the children as the still face in the original experiment.[14] "It's a real disruption in the attunement between parent and child," Dennis-Tiwary says. "When we're on our devices, not only are our heads down so there's no information socially, it's as if we disappear."

Work like this is a helpful reminder of *why* we need to be mindful of the effect being absorbed in our phones has on the people around us. It is unrealistic to tell parents never to use devices in front of their children or to expect the rest of us to gaze only at our companions. But the practice of requiring dinner guests to put their phones in a basket at the start of a meal is a healthy one. Eye contact matters. It triggers the social networks in our brains and that helps us maintain the strong, quality relationships we need.

Worry over disappearing into our devices has spurred some creative action. Filmmaker Tiffany Shlain is hardly antitechnology. She founded the Webby Awards and cofounded the International Academy of Digital Arts and Sciences. Yet for her own family she instituted what she calls a Technology Shabbat. In an echo of the Jewish practice of banning certain activities from sundown on Friday to sundown on Saturday, Shlain and her family refrain from the use of screens for twenty-four hours every weekend. Her family's Shabbat, she says, is about "being present" with the people in front of them.

This is one example of an adaptation to life with screens that helps to protect relationships. It's the kind of thing sociologist Claude Fischer of the University of California, Berkeley might have had in mind when he noted that people adapt themselves when they cannot change their environments, doing whatever is necessary to stay close to friends.[15] Fischer has been studying social networks for decades and in 2010 he reviewed forty years of research in his book *Still Connected*. Technology, he acknowledged, was the major change across that span, and the jury was still out on its effects when his book went to press. But he concluded that "for all the 'lonely, friendless in America' chatter in the media, the evidence suggests that friendship was as healthy in America in the 2000s as in the 1970s." (There were differences in the ways people conducted those friendships—entertaining at home was less common, but more time was spent with friends outside the home.)[16] What Fischer emphasized was that "people protect their core relationships."

Remember Sylvia, the bereft baboon who worked to find new grooming partners? You could argue that Sylvia was focusing on her core relationships—in her case, she was trying to reconstitute them, but she knew she needed them. Grooming is the primary tool that monkeys, like baboons, can use in their effort to strengthen bonds. Apes, like chimpanzees, have a slightly larger toolbox—in addition to grooming, they can share food and hunt together. And humans have the largest, most complex toolbox of all. Now we have added social media and digital technology to everything that was already there. While technology has changed some of the mechanisms we use to form and maintain friendships, it hasn't changed the role that friendships play in our lives.

I can't leave the topic of social media without talking about teenagers specifically. They are the source of so much of our anxiety in this regard. They are learning and growing, and their brains, as we have seen, are still working out the finer points of how to handle their social lives. They are a bit raw in every sense of the word when it comes to the ups and downs of friendship. Unquestionably, they make mistakes.

They don't always get it right on social media. Visual cues are missing online, and the artificial sense of distance and anonymity can result in saying and doing things kids wouldn't dream of face-to-face. They get hurt and they inflict hurt. But that is due in part to adolescence itself. "The way that young people are connecting digitally mirrors the way that they connect offline," psychologist Candice Odgers of the University of California, Irvine, says.[17]

In fact, the core qualities of friendship for teenagers are as present on Instagram and Snapchat as they are in the school cafeteria. Kids reveal themselves online (appropriately, more with their good friends than acquaintances). They find validation and support for everything from getting an assignment to advice on weathering their parents' divorce. They enjoy companionship through sharing jokes and memes and playing games together. And, of course, there is conflict online just as there is off-line. Mean girls and bullies were not invented with the iPhone, although their reach has expanded exponentially. There is also conflict resolution online, though it hasn't been much studied. Emoticons, for instance, help to clarify meaning in the absence of nonverbal cues.[18]

I see much of this at my house on an average afternoon. We live just a few blocks from my boys' school. If they don't have sports practice, the boys head home and hit the couch—an enormous leather sectional chosen for its ability to accommodate large groups (and to be forgiving of their spills and crumbs). Almost always there is at least one friend in tow. Sometimes there are four or more. They all sit down and start playing a video game. The preferred game itself changes year to year—*NBA2K*, *FIFA*, *Fortnite*—but the spirit of the thing does not. When playing together in person, they battle in pairs. If no friends are over, my boys use a headset to talk to friends who are playing with them on-screen—some of the friends live on the same block (as I write this, Alex is on that couch next door to my office playing with his friend Charlie four doors down), some of the friends live across town, and sometimes I find the boys playing with their cousins in Oregon.

In their enjoyment of video games, my boys are utterly typical. Video games, social media, and mobile phones play an integral role in how teens meet and interact with friends. In a large 2015 survey of teenagers, technology, and friendship, Pew found that more than half have made a new friend online (boys are more likely to do this than girls); 88 percent text friends at least occasionally and 55 percent do it daily. Nearly three-quarters of teens spend time with friends on social media; one-quarter do it daily. But—and this is important—the main place they interact with friends is in school, in person, and nearly all of them (95 percent) spend time with friends in person outside of school occasionally; 25 percent do it daily.[19]

For boys, video games play a critical role in the development and maintenance of friendships. In 2018, 97 percent of boys were playing video games. Like my kids, they rarely play alone. Such play has become a vehicle through which boys form and maintain friendships. More than a third of boys share their gaming handle as one of the first three pieces of information exchanged when they meet a potential new friend. (Only 7 percent of girls did the same.) Voice connections are used "to engage in collaboration, conversation and trash-talking," according to the Pew report. Among boys who play games with others, fully 71 percent use voice connections to engage with other players. More than three-quarters of online gamers said they feel more connected to friends they already know when they play games online.

For girls, texting represents the major change. While for all teens, texting is the default communication of choice, girls are heavier users than boys. Teenagers send an average of sixty texts a day, but one fourteen-year-old girl told the *New York Times* that she thumbs through nine hundred digital conversations a day.[20] There is some reason to worry about girls a little more than boys. In one of her 2019 studies, Amy Orben found a small but statistically significant indication that girls were more likely to suffer declines in life satisfaction from their time on social media.[21] Why that is remains to be seen and Orben's

next set of studies will investigate individual differences. It could be
that girls are a little more likely to suffer from depression and anxiety
than boys. It is also likely that societal pressure to look good in every
photo falls more heavily on girls than boys. I recently heard a story
about a group of college students who spent a great day out swimming
and having a picnic. On the way home, however, the three girls in the
group spent an hour arguing over which photographs from the outing
could be posted. One of those girls had an eating disorder. Here was an
example of the exacerbation of an underlying problem and the two-way
street that exists with social media.

It is fair to ask of all adolescents: What would they be doing if they
weren't online? We don't have any evidence yet to show that video
game interaction or texting is as valuable as face-to-face conversation, and plenty to suggest that it isn't. Work in this area is still under
way, but one of Ariel Shensa's recent studies touched on this question
when it found a connection between real-life closeness and depressive symptoms. Young adults who had a larger percentage of real-life
friends on social media, meaning greater overlap, were less likely to
have depression. "If we use social media as a tool to extend in-person
social relationships, great," Shensa says. "It's just another way to connect. However, having that kind of exposure to strangers could pose a
risk factor."

Adult fears about our children's time online is one of the areas where
we would do well to take a step back and look more carefully at what
we are objecting to—Is there friendship there, as there was with my
son Jake and his friend Christian when I got home from Puerto Rico?
Or is a child only ever playing games alone? Does a fifteen-year-old
girl spend the morning texting but then meet those same friends at the
mall or on the soccer field? Or does she stay home passively scrolling
through her phone? The difference could be significant. Pediatricians
are beginning to see troublesome social media use as part of larger
syndromes rather than as a problem by itself.[22]

Even for kids whose use falls in the normal range, adults don't always like what we see because we don't recognize it. I sometimes catch myself complaining that kids today don't seem to "talk" to their friends. For them, a text equals a conversation. Then I think back on my life as a teenager. My friends and I spent too many hours on the phone every evening, so much so that parents with the resources sometimes got a second phone line so as to be able to receive calls themselves. Be careful what you wish for, I'm sure my father would tell me.

Common sense is required both from parents and their offspring. It is helpful that researchers are beginning to establish a threshold for time on social media. Przybylski has developed what he calls the Goldilocks hypothesis: the idea that there might be a "just right" level of engagement with digital technology. "Moderate use is not intrinsically harmful and may be advantageous in a connected world," he and a colleague wrote. (Moderate looks like under two hours a day generally speaking, at least on weekdays.)[23] Other researchers are finding something similar.

All of us would do well to remember this: Social media is new, friendship is old—as old as life on the African savannas—and it is not going away anytime soon.

⤿

Reading all this research about social media and digging into Tanja Hollander's project got me thinking. One night I decided to do a little investigating. I went into my personal Facebook profile (I also have an author page) on which, at the time, I had 793 friends.

Who were those 793 people? Were they really my friends? I decided to go through the names one by one and categorize them according to how we knew each other. If I didn't instantly recognize the name or remember how we knew each other, that person went into a category called "question marks." Then later I went back to see whether I could assign them. Here's how it went.

Brooklyn and my children's school: 274 people

College: 178

Professional life: 109

Fire Island (where I spent summer vacations for fifteen years): 57

High school and childhood (including friends of my parents): 62

Hong Kong: 29

Relatives: 26

Miscellaneous (my husband's colleagues, former babysitters, and
others): 47

Question marks: 11 (Of these, only two remained once I looked at
their profiles.)

I was pleased to discover that there weren't more question marks. I was alarmed to discover that two people were dead, though their profiles lived on.

Seven hundred ninety-three is a lot of people. Clearly, my friendship style is acquisitive. My relationships—with both friends and acquaintances—persist. There are some other reasons these numbers are high: I have lived in my neighborhood for twenty years and have been very involved in my community; through my children, I'm connected to three classes of parents; my husband and I went to the same university (we share nearly two hundred friends); and my profession demands a public presence (I joined Facebook when my first book was published).

Some of the 793 are good friends, but three of my very closest friends are not on Facebook and others are not very active. More of the 793 are not really friends in that I rarely, if ever, see them in person, and our only connection is via Facebook. An interesting thing happened though while I was working on this book. My thirtieth college reunion occurred. I was excited to spend the weekend with my former roommates, like Sara, who gave me back the teddy bear, and women like Leah and Suzanne, who had recently become closer friends again after we became weekly accountability buddies and they weathered the ups and downs of writing this book with me. And there were a host of other former roommates (one year I lived with nine other women!) and good friends. Most of us do not live near one another and the prospect of a weekend in one another's company was delicious. We would drink bad beer and dance until the bands stopped playing and reminisce and revel in being together.

But also, I was looking forward to seeing a handful of people from my class I'd been following on Facebook. We weren't close, and probably never would be. Often, we lived thousands of miles apart. Our lives were busy. Our inner circles were full up. Yet because we were friends on Facebook, I knew about their lives a little bit. One was an acerbic political commentator who made me laugh on a regular basis. Another had recently weathered the hurricanes that devastated Puerto Rico. I knew about my classmates' children, their jobs, their passions for running marathons. When I saw those people in person, beer in hand under the reunion tent, we were able to skip over the usual repetitive small talk and get to more meaningful conversation.

"I'm glad to see you," I would say. And I really was.

EIGHT

Born to Be Friendly?

Perhaps I have an acquisitive friendship style because I was raised to entertain. Inviting people over was one of my mother's great joys. Her gatherings were unfussy and informal—a last-minute call to see who was free, or communally prepared meals over weekends in the country. The emphasis was on friends, food and wine, and conversation—the more political, the better.

From the time I was nineteen, many of those gatherings took place at a beautiful historic farmhouse called Compton, on the Eastern Shore of Maryland. The arrangement at Compton was unusual. My parents rented it together with several of their closest friends, but that did not mean the families divided their time there. The three couples that shared it spent their weekends together sailing, reading, and going for long walks through the fields.

My brother Mike and I were part of that group, treated like friends, and so was Mark, who first arrived at Compton as my boyfriend and whom I married under the magnolia tree in the garden.

"Who should we have for dinner?" Mom would ask, when she knew Mark and I would be coming for the weekend. And then, "What should we make?" She and I and whichever friends were on hand would go through cookbooks, run to the grocery store together, and then cook

up a feast, and my father would pop his head into the kitchen periodically to see if anyone wanted to go sailing. (When it got cold, he
resorted to enticing us with a flask filled with Southern Comfort and
vermouth—a drink he dubbed "Crew Sustainer.")

My mother has severe dementia now and has not been able—as I
write—to cook or carry on a conversation, for years. Yet well into the
progression of her disease, she would turn to me and ask, "Who should
we have for dinner?" and "What should we make?"

Her entertaining gene, we joked—because you have to laugh or you
cry—was going to be the last thing to go. Her drive to sociability felt
like it was hardwired into her DNA.

Even though I am my mother's daughter and throw similar parties,
albeit less frequently, there is, of course, no such thing as an entertaining gene, passed down with my mother's blue eyes and Irish complex-
ion. And even if there were, I never got the sense that I'm descended
from generations of hostesses. Certainly, my mother didn't grow up
learning from her own mother the way I did. Although there are stories of some extravagant childhood birthday parties, that all came to a
screeching halt when my grandmother, Ann, died of cancer when my
mother and her two younger sisters were just fourteen, ten, and nine.
My grandfather went to pieces after he lost his wife. There were no
more parties.

I always assumed the form of my mother's adult socializing was
largely cultural—a product of the middle-class, urban, professional
world in which my parents operated. My husband's experience was different. Mark grew up in a more rural part of Pennsylvania in a family
of steel and railroad workers where socializing was a family affair con-
sisting of grandparents, aunts, uncles, and cousins, all of whom lived
within a few miles.

And yet, over the last decade, scientists have discovered that there
is a kernel of truth to my joke about my mother's genetic wiring. There
is a biology to being social that includes genes. Dinner parties as a
way of socializing may owe something to socioeconomics and Western

tradition, but the underlying enjoyment of other people does not, at least not entirely. A tendency toward sociability is heritable, passed from generation to generation. And that confirms that friendship—born of a drive to connect and form strong bonds—is a product of evolution because genes are the field on which natural selection plays.

Now, it's a whole lot more complicated than saying I inherited my friendliness from my mother. Just as there is no such thing as an entertaining gene, neither is there a gene for friendship or for any other complex trait or behavior. Or to put it another way, any one gene can explain only an infinitesimal amount of the variation between me and my much more introverted friend Elizabeth. The last twenty years have also brought a far more sophisticated understanding of the way genes and environment interact, a field of inquiry known as epigenetics. Even if there were a gene for friendship and I carried it, I still might not be all that social if my lived experience squashed the expression of that gene.

Nonetheless, the two main thrusts of genetic analysis have a lot of potential when applied to social behavior. Genetics can help explain human universals, such as the larger drive to cooperate. And they might also tease out some of the variation in behavior from individual to individual—why one person is shy and another outgoing. (Differences between groups tend to be cultural.) As the importance of social relationships becomes ever clearer, it is logical to begin looking for further answers in the genome and that is just what scientists studying friendship—and its effects—are doing.

"Genetics started with an understanding of how genes affect the structure and function of our bodies and then our minds, and now people like us are beginning to ask how genes affect the structure and function of our societies," says sociologist Nicholas Christakis, whose work on social networks was featured in chapter 6. Like my mother, Christakis likes dinner parties. "My favorite thing to do in the world is have two couples that we like come over and have a dinner party for six people," he says.[1] In such an experience, he sees pieces of what makes friendship feel good—the sustained relationships, the reciprocity (presumably those same friends will have you over in the future), the positive emotions that flow with the wine. But Christakis's work on networks is less about the dynamics around one table of six and more like understanding the dynamics at a gala dinner for one thousand or even ten thousand if one could find a dining hall large enough to accommodate such a crowd.

As he and James Fowler waded further into the question of what underpinned social network formation, they were interested in the effect of individual characteristics—what made the structure of my social network different from my neighbor's or from Christakis's, for that matter. The models designed to explain how networks emerge—and there are many—had, to that point, mostly ignored what part individual differences play. Thinking about individual differences leads fairly naturally to thinking about genes and to applying the ideas of evolutionary biology. When Christakis and Fowler set out to explore the possibility that humans are endowed with traits that affect their network attributes, they had to reckon with the fact that "our most intrinsic characteristics can be found in our genes."

If you want to tease out the role of heredity in any particular trait, an obvious place to look is twins. Even before scientists fully understood heredity, they recognized that twins would be a good way to study inheritance. In the 1920s, when it was determined that identical and fraternal twins are genetically different, the idea really took hold.[2] Identical twins are the result of one zygote that splits and forms two embryos, whereas fraternal twins begin as two separately fertilized eggs. As a result, identical twins share 100 percent of their DNA and fraternal twins share only about 50 percent of their genes, as much as any other pair of siblings. If genetic variation contributes significantly to a trait, then identical twins should be more similar in that trait than fraternal twins. Moreover, twins can also provide a lens through which to examine the effects of environment—as in the case of fraternal twins raised in the same home or identical twins raised separately. This approach has been used to show that genes play a role in personality, intelligence, and other behavioral traits. (A small note of personal complaint: my brother Michael and I are twins, and I confess amazement at the number of people who don't immediately realize that because we are different sexes, we cannot be identical. It seems that very few people have thought through the genetics of twins.)

Twin studies helped in the development of what several writers have called one of the trickier concepts in modern biology: heritability. Some inherited traits, like eye color, depend purely on recessive and dominant genes. I have blue eyes (recessive) and my husband, Mark, has brown (dominant). But two of our three sons have blue eyes because Mark carries a recessive gene for blue eyes from his mother and that's the gene he passed on to Jake and Matthew. His brown eyes won out in Alex, our youngest son. Other traits, like height, are a little more variable. While a person's height is strongly correlated to the height of their parents, it depends on multiple genes and can also be affected by diet and environment. All three of our boys are six feet or taller while I am only five feet six inches and Mark is a shade under five feet eleven. (The explanation lies with the rest of my extended family. I am the shortest by far.) How to account for these kinds of differences in the way a trait manifests itself in later generations? The visible presentation of a trait or behavior (its phenotype) is partly due to underlying genetics (genotype) and partly to environment. Nature and nurture are always both at work, though their relative contributions vary depending on the circumstances. The fraction of variation in a trait that can be attributed to genes—that can be inherited—is known as heritability.

It was heritability that Christakis and Fowler were seeking. In search of genetic underpinnings to social networks, they turned first to a twin study. Conveniently, the National Longitudinal Study of Adolescent to Adult Health (known as Add Health) contained both genetic information and data on friendship. It also included 1,110 twins. Christakis and Fowler compared the identical twins to the fraternal twins genetically and socially. Their resulting paper, published in 2009 in *Proceedings of the National Academy of Sciences*, was the first to mix genetics and social networks—in any species.[3] In it, they calculated the heritability of several different individual properties essential to the creation of a human social network.

Bear with me for a short glossary of the terms of social network analysis. "Centrality" is what it sounds like, a measure of how well connected an individual is to others. "Strength" weights the relative importance of each connection. "Eigenvector centrality" measures popularity by counting the strength of your connections' connections. (In effect, it tells us that cool kids have cool friends.) "Transitivity" is the likelihood that any two of a person's contacts are connected to each other or also friends. "Degree" refers to the number of a person's contacts or social ties, counted both outward and inward. In-degree reflects how many times a person is named as a friend by others, and out-degree is the opposite: how many friends a person names (the two do not always align).

A map of such a network looks vaguely like a densely complicated Alexander Calder mobile. Lines and circles lead to more lines and circles, like so many three-dimensional hubs and spokes. Each circle represents an individual and the lines are the connections between individuals. Someone at the center of a social network will have many lines connecting to his or her spot on the map. A loner on the periphery will have very few. These two individuals are like the difference between the Times Square subway station in New York City, where eleven lines converge, and a local stop in an outer borough served by only one train line.

Christakis and Fowler found that some but not all network characteristics were heritable. Genetic factors accounted for nearly half of the variation in how connected an individual is to a larger friend group (based on the number of in-degree and out-degree associations). In other words, there is heritability to how many friends people have. Like height and other traits, your friendliness varies. That's not a big surprise to most people because we know that some people are born shy and some people are born outgoing. People vary in their appetite for friendship.

More surprisingly, they found that genetics also explained just about the same amount of variation in transitivity—the probability that a person's friends are friends with one another. "That's a bizarre result," Christakis says. "If you have Tom, Dick and Harry in a room, whether Dick is friends with Harry depends not only on Dick's genes or on Harry's genes but on Tom's genes. How can that be? We think that the reason is that people vary in their tendency to introduce their friends to one another. Some knit the networks around them together, and some people keep their friends apart."

Finally, they found that 29 percent of a person's centrality, whether they are at the middle or the periphery of the network, can also be explained by genes. Christakis can only speculate as to why that would be so, but he ventures to guess that what's happening is that people differ in their taste for popular friends. "You might want four unpopular friends, so you can have their undivided attention. And I might want the same number of friends, but I might want my friends to be popular because I want them to be well-connected. Even though I don't get much of their attention, they in turn have many connections themselves. These are different strategies." The upshot of all of this is that the variation across individuals in their propensity to introduce their friends to one another, and in their propensity to pick popular and unpopular people as friends, comes to be reflected in the actual structure of the social networks we humans construct. Therefore, that structure depends in part on our genes.

In 2011, Christakis and Fowler went further. They wanted to test for genetic similarity among friends who had no biological connection. To do that, they examined six available genotypes from the same database, the Add Health study, but this time, they excluded relatives. Their results suggest that the old adage about "birds of a feather" has some basis in genetics. Friends did not just have similar traits. They resembled one another on a genotypic level.[4] The results went well beyond what one would expect from systematic genetics differences that might occur because of shared ancestry, such as being of European or Asian descent. In other words, it isn't just that we hang out with people who look like us. Christakis and Fowler expanded on this work in a 2014 paper on friendship and natural selection, again published in *PNAS*, in which they showed a degree of correlation in genotypes that made friends the equivalent of fourth cousins.[5] That held true for the Add Health database and also the much larger database from the Framingham Heart Study. "Friends may be a kind of 'functional kin,'" they concluded. It remains a mystery how we recognize such "functional kin" in order to befriend them.

On Cayo Santiago, the island off the coast of Puerto Rico where we began this story, Michael Platt and Lauren Brent are applying social network analysis to groups of rhesus macaques. They were also among the first other researchers to follow Christakis and Fowler into the exploration of heritability of social behavior.

Platt learned his craft as a graduate student in biological anthropology in the lab of Robert Seyfarth and Dorothy Cheney at the University of Pennsylvania, so a keen interest in social behavior is part of his academic inheritance. "What could be more interesting or compelling than understanding how our brains and bodies come together to form relationships—friendships, alliances, lovers, enemies, societies?" he once said. "And what could be more relevant than deciphering how these relationships fall apart or fail to form in the first place, and how to mend these rifts?"[6] After a long stint at Duke University, Platt returned to Penn as a professor in 2015, which is where I met him for the first time. Like Christakis, Platt teems with ideas and energy and amazes those who work for him with the breadth and depth of his knowledge. He is officially a neurobiologist, yet his work encompasses everything from autism to decision making to what makes us human; it spans anthropology, primatology, and neuroeconomics. But the theme of sociality—and, by extension, friendship—runs through everything he does.

On Cayo, Platt can get at individual differences in social behavior and their possible biological causes. The list of possibilities is long: sex and age, physiology, genes, neural architecture, social structure, and finally, reproductive success and evolutionary fitness. In most other social animals, from fish to squirrels to dolphins and whales, the behavior you can observe is limited, but rhesus macaques present a microcosm of primate society that is as varied as any large human group. Some of the monkeys are deeply embedded in the social world, others are more peripheral. (Think back to ninth grade. Didn't your class include the quiet and the boisterous? The kind and the mean? Academics and jocks?) Why such variation? In addition to offering large numbers of animals living and socializing freely as well as reams of data on them going back for generations, Cayo also has genetic information, obtained once a year when the animals are trapped so that blood samples can be drawn. That makes it possible to look at how genes interact with environment to shape behavior. "Whatever animals do on Cayo has an impact on their success, on their ability to survive, and thrive and mate and have offspring," says Platt. "In a sense, we can see evolution happening."[7] His ambition is to do social primatology on a massive scale. Sitting on a park bench in Philadelphia, far from Puerto Rico, he grins. "Let's watch all the monkeys."

To get the project under way, while he was still at Duke, Platt hired Lauren Brent in 2010 as a postdoctoral fellow. She had already spent a year on the island doing research for her dissertation in which she linked stress responses with monkeys' social relationships. She theorized that if having stronger social bonds is adaptive, animals more central to their social network, and possibly those with more indirect connections (friends of friends) should show tempered stress responses. Brent built a social network map of all the animals in her study group, and measured stress hormone levels in fecal matter, from which it had just become possible to pull enough hormonal information to perform meaningful calculations. It turned out that

high-ranking females with a lot of indirect connections did indeed
have lower stress hormone levels.

Following Jeanne Altmann's protocol, Platt and Brent chose fifty-
five behaviors—feeding chow, feeding water, traveling (on the move),
giving grooming, receiving grooming, resting, fighting, and so on—
all of which would be entered into the Psion handheld computers the
research assistants use. Each monkey is followed for a total of twenty
minutes per week. That may not sound like much, but it adds up. After
the first five years of their project, they had hours of data on 450 animals, and DNA on 900 animals. "The goal is to completely character-
ize every animal in the population and assess the continuity of their
behavioral characteristics over time. And then as completely as possible assess their biology," Platt says.

One rainy morning on Cayo, I followed Aparna Chandrashekar, a
recent college graduate hired for a yearlong stint as a research assistant,
to see what focal sampling on Cayo entails. The rain had made the tree
roots slippery and a bit treacherous, and hermit crabs had emerged
seemingly from nowhere.

This was something of a dream job for Chandrashekar, who had
monkeys on her bedsheets as a kid and whose favorite food was bananas
because of Curious George. Now, real monkeys filled her days. She had
to sample each of the fifty-seven adults in her assigned group twice a
week. On her best days, she got through thirty—fifteen in the morning
and fifteen in the afternoon. She began each morning with a table of six
time blocks and a list of which of her animals were eligible for following
that day—some days, there were a lot, other days just a few remained.
Focal sampling is an endlessly repeating process of elimination.

No sooner had we stepped out of the fenced-in cage where we left
our lunches and backpacks than we found one of the monkeys on her
list, an elderly male, 94Z, sitting outside the feeding corral. He had a
long, triangular face and a bit of a hunchback walk. Chandrashekar
entered his coded details in her handheld computer. For the moment,

she showed him at rest. Over the course of the next ten minutes, she recorded that he ate some monkey chow, and that he was being vigilant, looking around constantly. He was an interesting case because he was once the alpha male in another group but lost his power and moved into this group. Perhaps that was because his leadership style was fairly laissez-faire, says Chandrashekar. "I've never seen him give aggression."

That said, he did occasionally remind the youngsters of his status. "He just stare-slapped that juvenile," she pointed out after he hit the ground and fiercely ogled a young animal nearby.

Ten minutes were up, and we moved on and found three monkeys lying together on the corrugated roof of the enclosure we had just left.

"They're giving themselves back massages," Chandrashekar pointed out as the animals rolled from side to side over the metal bumps.

It had rained earlier, and water was dripping from the roof. One monkey leaned down to drink. Another appeared to think that was a good trick and did the same.

"They like the rain. It's playtime."

But these easy-to-find monkeys were not on her list for the morning. Instead, we were searching for 1G9, a loner. He was nowhere to be found. Because of his tendency to disappear—and a short stint in a different group—the researchers were behind in recording observations on 1G9. In order to catch up, Chandrashekar had to find him every day.

"He's usually back here," Chandrashekar told me as we crouched down to search under an expanse of bushes well away from the main group.

"He's a big guy but he definitely likes to hide."

We questioned another researcher as we passed by.

"Have you seen 1G9? He has a patch of fur missing on his back."

No luck. We came at the bushes from the other side and lay down on the ground so as to see a little farther.

"I always find him in here," she said. "Maybe someone chased him off. He'll probably come back. That's how it goes."

Frustrated, she moved on to another focal sampling, but she would have to look for 1G9 again later.

These are the minute-by-minute on-the-ground observations that will paint, when this project is done, a much bigger picture of what it means to be social. On Fridays, at the end of a week of collecting data, the researchers devote the afternoon to uploading their data into the software program back in the office in Punta Santiago. If they know they made a mistake—like entering an approach when they meant to enter a displacement—they make a note and correct it on Friday.

At the end of each year, they create a master file of that year's data from which Brent compiles extensive coded spreadsheets, just as Joan Silk did with the baboons of Amboseli and Moremi. There's some cleaning up of the data to do and a few decisions to make. But when Brent has it assembled, she has graphic representations of detailed behavioral observations like the snippet of 94Z's morning that I just watched. Out pop charts of Grooming Received for 2015 or Aggression Given for 2016. (When she e-mails the resulting graphs to the field team so they can see the results of their labor, she calls it "Monkey Christmas.") She also creates networks of proximity, maps of who hung out near whom. Proximity may seem underwhelming as a social behavior. Does it really matter who is nearby if the two are not directly interacting? In nonhuman primates, the answer is yes. Monkeys and apes don't tolerate strangers, so proximity signals a positive bond. That is why primatologists joke about what would happen if anyone ever loaded a group of unfamiliar chimpanzees onto a plane together and expected them to sit quietly for a flight of any duration—something we humans manage thousands of times per day.[8] It seemed likely that when 1G9 eventually showed up as a dot on one of Brent's social network maps, he would be a lonely dot. Sure enough, two years later, the record showed that he did not groom with any other animal in the time Chandrashekar was watching him.[9]

Social networks alone revealed several interesting things about the makings of social behavior. Since female macaques stay in the groups into which they are born, logic suggests that older females who have developed more social skills or just know everybody better would be the most socially central. That's exactly what Brent found. Females were playing a much bigger role in the social networks than males. They give and receive more grooming or proximity and older ones receive less aggression. "This is either because they are more socially skilled and know how to stay out of trouble, or because they have less energy and are less in the mix," she says.

The extensive data have allowed Brent to build a dominance hierarchy for every group. The lists are based on recorded wins and losses during interactions between individuals. "If I were to walk up to you and you moved, and I took your chair, then I win, and you lose," Brent explains, describing what they call a "displacement." There are also ritualized submission gestures—the fear grimace or the silent bared teeth display, which looks like a smile but means *I submit, don't hurt me*. "Or they'll just lean away or even crouch down or lie down on the ground like a dog would do," says Brent. "If you do that to somebody that's a clear loss."[10] The resulting charts list one monkey after another in descending order of status, like tennis rankings in which players hold their spots for years. But dominance hierarchies didn't explain the whole social network at all. High-status individuals are often the best connected, but not always. And so, the Cayo scientists turned to genes, asking whether there was a genetic signature to where a monkey sat in the social network.

If you want to understand the genetics underlying behavior, there are several possible approaches. The mapping of the complete human genome in 2003 made possible genome-wide association studies (GWAS), which compare an individual's entire genome against those of many others in search of anomalies. If someone with Parkinson's disease has variations in particular genes that aren't found in people without Parkinson's, for example, those genes can be said to be associated with the disease. (Whether or not they can be said to cause it requires more work.) The technique is continuing to improve,[11] but early enthusiasm has died down some. GWAS has proven useful in diseases that are tied to one or a handful of gene variants, such as age-related macular degeneration. But to use genome-wide association studies to reveal genetic variation in more complex traits or diseases, you need thousands or even millions of subjects for comparison. After a decade of effort, a 2017 GWAS study of height used the genomes of more than 700,000 people and turned up almost eight hundred genes influencing how tall someone grew to be. Together they accounted for just over 27 percent of the heritability of height. The researchers involved, who had been laboring at this for years and initially found only one gene they could reliably tie to height, were understandably pleased. But others thought the results were disappointing, not to mention a waste of money, since they left so much about height unaccounted for.[12] Whether you saw the genetic glass as half full or half empty depended entirely on your perspective.

A second approach is to pick candidate genes that you have reason to believe play a role in the disease or behavior you're studying. You have to know going in that the genes you are looking for have something to do with the trait that interests you.

This was the approach Platt and Brent and their colleagues adopted on Cayo. They decided to look in the serotonin pathway, which is important in the brain for social reward and therefore for social relationships. In humans, one of the alleles they studied, *5-HTTLPR*, has been linked to elevated levels of anxiety, elevated response to social threat, and a tendency to avoid risks. The other, *TPH2*, has been associated with autism and depression. "If there are variants in the serotonin pathway so that you have a lower functioning pathway then maybe you're going to be a really asocial monkey," Brent explains.

Platt and Brent found, in an echo of Christakis and Fowler's work, that social network position is heritable in macaques. More closely related animals were more similar in centrality than those that were more distantly related or unrelated. Social behaviors over which animals had some control, like giving grooming, being aggressive, or spending lots of time in close proximity to others, were most likely to be heritable. Rhesus macaques carrying the less frequent variants of the serotonin genes they analyzed tended to have fewer friends and allies, probably due to either poor social skill or lack of interest in others. In their 2013 paper in *Scientific Reports*, the researchers concluded: "Together, these results endorse the idea that the capacity to form social relationships mediated by other individuals has been shaped by natural selection and may play a crucial role in the evolution of primate societies."[13]

Platt and Brent have tied variation in genes that in humans are associated with differences in behaviors to genes in monkeys that seem to predict different kinds of social outcomes. That puts the group on the road to creating monkey models of social functions—reward, communication, and so on—that are compromised in people with autism and other disorders. And, of course, it tells us more about what underlies the basics of friendship. It looks like the monkeys that are central to their networks and have lots of grooming partners are more likely to differentiate between individuals in the group and out of the group. "They just hold the big picture," Platt says. "Maybe . . . that's a hint that individuals who are at the center of the network have more social skill and more social attentiveness."

The Cayo researchers also followed the example of the baboon watchers in Africa by measuring reproductive success in light of their social network and genomic data. "If being really social or being really aggressive has been selected, then we expect that the animals that do that would have greater reproductive output," Brent says. Behaviors that show up in the most social monkeys—receiving more grooming, hanging out in close proximity, being central in the network—did indeed lead to more babies.

⌐

Far from Cayo, a small female rhesus macaque hangs almost horizontally on a chain-link fence looking at me. Her head is cocked to one side and she pulls back her lips, showing me her teeth. "There's the lip smack she's doing for you," primatologist John Capitanio of the University of California, Davis says. The lip smack accomplishes several things. It signals a little fear and submission, but also a desire to affiliate. He points at the way she's displaying her body. "That's almost a groom present right there. Come and groom me."

I can't. Not just because it wouldn't be safe, but because Capitanio and I are not just on the outside of the fence but sitting in a car. Capitanio is giving me a tour of the California National Primate Research Center. Located a few miles outside of Davis in the middle of a run of farm fields, the primate center is one of only seven in the country. About forty-five hundred monkeys, nearly all of them rhesus macaques, live here. As on Cayo, these monkeys need to be protected from humans like me who haven't been medically cleared. Most of the animals inhabit half-acre field cages like the one we are parked beside. The enclosures are full of hand-me-down playground equipment— plastic slides and climbing towers, ropes to swing on, and barrels to roll. Each cage is home to between fifty and one hundred fifty animals, organized like a troop in the wild along matrilineal lines. But there are also indoor laboratories and a set of repurposed corncribs used as transitional housing depending on what the studies under way require.

Capitanio has been watching rhesus macaques for forty years, much of that time here at Davis, where he got his PhD and is now director of the Neuroscience and Behavior unit. Lanky and long-haired, Capitanio is an unabashed graying hippie. He almost didn't go to the long-ago abnormal psychology class at the University of Massachusetts Amherst that turned him on to studying animal behavior. It was a beautiful day, he had a joint in his pocket, and there was a good band playing. But Catholic guilt pushed him inside where the guest lecturer was a psychologist who had been Harry Harlow's last student. Capitanio was hooked by her talk, became an assistant in her lab, and has since been following these animals as devotedly as he ever followed the Grateful Dead.[14]

The behavior of the female we are watching, her bold yet respectful seeking of connection, is telling. Not every monkey does that, just as not every human approaches a stranger. Over the years, Capitanio has become a connoisseur of the big-picture distinctions and finer-grained nuances in monkey social behavior, and he is convinced that they have a lot to tell us about humans.

Comparative studies across species force you to strip behavior down to its simplest forms, and Capitanio works from the simplest definition of friendship I have run into so far. Two animals are friends by his calculation if they hang out together happily, without conflict or aggression, more often than chance would predict. Such a mathematical approach does not begin to capture what we mean when we talk about friendship in humans, and Capitanio is quick to acknowledge that, using as an example his relationship with his oldest friend, with whom he has a long history, deep understanding of shared backgrounds, and the ability to pick up where they left off. "I'm not saying that a friendship between two monkeys is the same in all of that regard," he says. "The question becomes, if it's not exactly the same kind of thing is it useful? And my response is, hell yeah, it's useful."

There are two main arguments as to why it is useful: common ancestry and homology. Rhesus macaques and humans diverged about twenty-three million years ago. Capitanio likes to feature a shared ancestor called *Aegyptopithecus* in his lectures. Its name means "linking Egyptian ape" and based on the fossil record, Capitanio says, "it lived 24/7 in a social group kind of like this although probably smaller." Similarities that exist between species because of such a common ancestor are called homologies. "Homologies are great things because what that suggests is that the probability of our finding something in your brain that's similar to something in her brain"—Capitanio points to a female macaque—"are much higher in a homologous situation than in the opposite, an analogous situation." And because scientists think there are homologous structures in my brain and the brain of the monkey we were watching, the likelihood of learning things relevant to humans from manipulating the social lives of these monkeys is much higher. Capitanio is also excited that the work he is doing in the lab has been validated by similar findings in the field, such as among the baboons in Amboseli and Moremi. "Different species, different scientists, different methodology, but what do you know, social integration seems to be good for your health."

A place like Davis allows experiments that would be impossible in the field or with humans. Even so, some of what Capitanio does consists of the same kind of observation done in the wild. He distinguishes social inclinations by watching each of the possible repertoire of social behaviors monkeys engage in. "Grooming is gold." Based on observations, he and his researchers tally up a sort of social score for each monkey. How many approaches did an animal engage in? How often did it pay off?

What Capitanio has found is that each troop of monkeys divides fairly evenly according to temperament. One-third are highly social. Engaged and successful, they seek social interaction and they get it. Another third of the group is average. And the final third is what Capitanio calls *low social*. They are much less connected to the rest of the group. Except along the way, Capitanio noticed a wrinkle. While all high-social animals were roughly the same, low-social animals fell into one of two categories. Some seemed to be introverted. They did not engage in much socializing, but neither did they seek it out; they conducted very few approaches or walk-bys. The other low-social animals were different. They tried to make friends. They tried to hang out. They just didn't often succeed, and their efforts did not convert into connection.[15]

That made these monkeys of great interest to John Cacioppo, the loneliness researcher from the University of Chicago, when he began to wonder whether he could create an animal model of loneliness. After all, you can't just give monkeys the UCLA Loneliness Scale. And furthermore, if you could create a monkey model of loneliness, what would it tell us about the biology of loneliness that is harder to see in humans?

When Cacioppo teamed up with UCLA genomicist Steve Cole and published that first surprising paper showing clear changes in immune system regulation in lonely people, Cole's years of studying HIV helped him read the tea leaves of gene expression easily. He was blown away by what he saw. "When we looked at the genes that were under-active in the white blood cells of lonely people, it was chock full of genes involved in that antiviral response that I just happened to have been studying for the last ten years," Cole says. "I looked at all those things and I was like, Holy Cow! No wonder these people get sick more often. This is a recipe for disease."[16] In short order, they replicated those first findings in much larger groups of Cacioppo's study participants, first 93 and then 141.[17]

It was Cole who connected Cacioppo to Capitanio. Cole and Capitanio had already collaborated on Cole's HIV work[18] using the monkey version of HIV, which is called SIV, for simian immunodeficiency virus. Although you certainly cannot re-create the experience of a closeted gay man in a rhesus macaque, you can create what Cole calls "a reasonable facsimile for the chronic sense of threat that seemed to be the driving psychology of closeting." Capitanio did this by housing monkeys in different kinds of social groups—stable or unstable. On day one of the experiment, he took animals out of their cages and put them together with unfamiliar animals for one hundred minutes. Then on the subsequent day he did it again, but some got put back with the same new animals every day, giving them time to get used to one another, and others were placed with a new group each time. If you want to stress out a monkey, putting it with a different set of animals every day is a good way to do it. "From day to day we would just mix them up," says Capitanio. "The animals in the stable and unstable social conditions have the same amount of social opportunity, but the stable social situation permits the development of more elaborated and deep relationships." He and Cole found that this artificial social stress in the unstable condition caused the SIV in those monkeys to replicate faster, just as it had in the closeted gay men Cole had studied.[19]

Then they did a further experiment. They removed a set of each monkey's lymph nodes, where the virus replicates, in order to look at the nerve fibers of the sympathetic nervous system. Those nerve fibers do their work primarily by releasing neurotransmitters, such as norepinephrine, which is triggered by the fight-or-flight stress response. Cole and Capitanio wanted to see whether the nerve fibers were next to cells that had SIV in them. To their surprise, they weren't just next to them—there was considerably more SIV where the nerve fibers were. "There are always nerve fibers in lymph nodes," Cole explains. "They tend to be wrapped around blood vessels and every once in a while, they'll send out a little branch here and there. But in the animals that were in these unstable social conditions for a couple weeks, those branches really arborized. They sent out tentacles all over the place."[20]

This was not at all what Cole had expected. He had been taught that nerve distributions were, by and large, static—they were set up by basic developmental programs and then sit there, leaking neurotransmitters. "This made it look like the nerves were actually pretty active," says Cole. "If you are stressed for a long time, those nerves are going to grow out more branches and basically create a bigger pipeline from the brain down to your immune system." It looked like they had uncovered at least one of the biological mechanisms that converts stress to disease. Those results matched up well with studies Cole had done years earlier, where he put human cells in a test tube, infected them with HIV, and added norepinephrine. Lo and behold, the HIV virus replicated faster.

As all this work was under way, basic laboratory science was beginning to appreciate that inflammation is like a generic fertilizer for just about every disease that afflicts us. HIV, Cole realized, had evolved to take advantage of our stress biology, which causes us to shut down our viral response and ramp up our defenses against bacteria in the face of stress. Once upon a time, that would have been the odds-on right response since the kinds of things that killed people were untreated wounds and infectious diseases. But today, in an era of chronic disease like metastatic cancer and heart disease, it's the opposite of what we would wish for. "Shutting down the human antiviral response? If you're HIV, this is a great time to wake up and replicate like crazy," says Cole. "Most viruses have learned to counterprogram against the human immune system by listening for stress biology and amping up their activity at the same time as the immune response against viruses is, for evolutionary reasons, shutting down."

The pivot that Cole had made was to recognize closeting in gay men as a primarily psychological event and to think about how that lived experience of the host influences what the virus does. That is exactly what Cacioppo wanted to know about loneliness. He saw it as a threatened psychological state. And their first gene expression study had revealed that loneliness was indeed showing up on a molecular level. If they could examine this in monkeys, they could see whether something similar was happening in the lymph nodes of lonely animals.

"What we've learned about genes is mostly that your genetics matter not because they determine for sure what you're going to be like but because they change your sensitivity to environmental influences," Cole says. "The genes lay out the basic architecture of a human being—a menu of options for creating human proteins. But almost everything in terms of who we actually become, which of those proteins actually get made, especially the later we go in our lives, is shaped predominantly by the environments that that genetic project is unfolding in. We're all a joint product of a particular human genome and a particular life." Which genes are on and which are off, then, makes all the difference in the world in how a body runs and how an individual behaves. Cole's work is illuminating the fact that our social environment is just as critical an element as the physical environment, the air we breathe, and the food we eat.

Together, Cole and the two Johns, Cacioppo and Capitanio, began exploring the biology underlying the differences between the two kinds of low-social monkeys Capitanio had previously identified. In the monkeys that seemed to yearn for more connection, just like in the adults from Cacioppo's Chicago study, there were clear signs of the same pattern of immune system dysregulation. Inflammatory genes were up-regulated and viral protection was down-regulated.[21] "We think lonely monkeys as well as lonely people are in this bind," says Capitanio. "They want more social interaction but they're also afraid of social initiations. They're afraid they're going to get rejected." That fear triggers overactivity in the sympathetic nervous system, which ramps up production of a subset of white blood cells called monocytes that are released from the bone marrow and that are primed to create inflammation.[22]

Intriguingly, Capitanio does not find the same kind of underlying physiological changes in the second group of low-social monkeys, those that don't try to interact with other animals very much. In fact, their biology looks very similar to that of the high-social animals. "I think that the critical distinction here is that there's some degree of 'satisfaction,'" says Capitanio, marking the word "satisfaction" in air quotes with his fingers. "I'm putting that in air quotes because I don't know that that's what's going on in the brain of the monkey."

That first paper on changes in gene expression in lonely people didn't just excite the public (remember, Cole received more e-mail about it than for any other study). It also brought a wave of other scientists to Cole's door. Mostly, these were scientists who were interested in social stress and intrigued by the prospects of seeing changes in the genome. They offered themselves up as collaborators with Cole. The first out of the gate was a friend named Greg Miller, a psychologist who studied how stress affects health. Miller, who is now at Northwestern University, had been studying older women caring for husbands dying of brain cancer. Like Cacioppo, he had a freezer full of the women's blood samples, so he and Cole did a quick assay.[23] "Damn if we didn't see essentially the same profile in the white blood cells," says Cole. "It wasn't all the same genes, but the signature of inflammation being

up-regulated in these older women taking care of their dying husbands and antiviral response being down-regulated—that signal came
through clear as a bell."

Cole began looking for the response in a range of people suffering
from various kinds of adversity. Again and again, there it was. In people
suffering from posttraumatic stress disorder. In people struggling with
poverty. In child soldiers from Nepal. In breast cancer patients. The
consistency of the response was so striking that eventually Cole gave
it a name: conserved transcriptional response to adversity (CTRA).[24]
The name may not come trippingly off the tongue, but it reflects the
essential elements of this process that he has uncovered. The term
"conserved" is used in biology to mean that a trait appears across species. That was true here. In addition to humans and macaques, CTRA
has been found in mice and even fish. But *conserved* does double duty
because this response appears across a range of risk factors as well.
What they all have in common is a sense of threat and uncertainty.
That threat activates fight-or-flight biology. "Transcriptional" refers to
the critical step of triggering a gene to turn on or off—its DNA must
be transcribed into RNA for the gene to have an effect on the body.

This echoing response to adversity did nothing to diminish Cole's
respect for the seriousness of loneliness, where he had first found it.
It only increased his appreciation for the deep importance of social
connection. "Quiet suffering may be how people experience loneliness,
but at the molecular level it ranks right up there with poverty, trauma,
bereavement, and all kinds of other much more vivid, dramatic stuff,"
he says. "Loneliness is one of the most effective ways we know to make
a body feel threatened and insecure."

NINE

Deeply Built into the Brain

The fMRI machine at Dartmouth College is tucked away in the basement of the psychology building. Outside, it's a hot August day, but the only signs of summer down here are the technician's shorts and Hawaiian shirt. We sit in the control room watching the machine in the next room on a large wall-mounted screen. The angle of the camera makes the tube look like something out of a rocket ship. It reminds me of science fiction movies featuring astronauts passing through narrow tunnels to move from place to place.

All we can see of the student volunteer lying in the scanner are his knees and feet. For privacy, I will call him Brad. Clad in hospital scrubs so as not to bring any metal into the machine, Brad's legs jut out of the magnet bore with his knees propped on a pillow for comfort. When he was getting settled into the foam headrest that holds his head still, I caught a glimpse of his dark hair and eyes and long narrow face.

Brad is twenty-eight years old and has just arrived on campus to begin a graduate program. This visit to the MRI lab is among the first things he has done since he got here, and that is the point: to catch him before he has had a chance to get to know anybody.

I have arrived—technologically speaking—at the cutting edge of the science of social behavior. Here in this basement lab neuroscientists

Thalia Wheatley of Dartmouth and Carolyn Parkinson, who was once a graduate student in Wheatley's lab and now runs her own lab at UCLA, are investigating what friendship looks like in the adult brain. Granted, Brad is alone in that machine, with no friends in sight. But Parkinson and Wheatley have discovered that what they see today in Brad's brain will allow them to make some predictions about the texture and tenor of the friendships he will make here on campus.

In several preceding chapters, we watched the social brain develop, first in infants and children and then in adolescents. Babies fixated on faces and their mothers' voices, loving caresses sent happy signals to their brains to encourage bonding. As they grew, those babies began to see the world through other people's eyes and ears, to realize that other people had minds of their own, filled with unique ideas, beliefs, and perspectives—a prerequisite to friendship. Through play and early school experiences, children learned to be part of the social push and pull of a group, to cooperate, to trust and be trustworthy—further steps on the road to friendship. As adolescents, they were hypersensitive to the rush of reward that comes from feeling connected and accepted, from having intimate friends to confide in and have fun with. Many of them also felt the crushing pain of exclusion when hurt or left out.

Eventually, those babies and teenagers grow up to be adults, like Brad. As we have just seen, each of us brings to adulthood our own genetic propensities to be more or less friendly, tendencies that have been tweaked by the people and places—the environment—in which we grew to maturity. But the relationship between genes and social behavior is under the control of the nervous system, of the brain and the signals it sends through our bodies.

An adult brain has years of experience processing social stimuli. Most grown-ups have mastered theory of mind. (If they haven't, that is a sign of a neurological disorder.)[1] The cortical brain regions that handle reason and judgment have caught up with their emotional counterparts in the limbic system and fully matured. Make no mistake, as adults, we are still driven by the neurological rewards of being social, but we are better able to weigh these rewards and have the capacity to use our inclination to cooperation and affiliation for the greater good.

As a result, we humans are remarkable for what our sociality makes possible. Even though monkeys and apes, dolphins, and other species can do much of what we can, they have yet to join together in the extreme forms of cooperation we have achieved. They have friends, but not like we do. Other animals are similar to humans in kind, but not in degree. Now that we have come to more fully appreciate the critical importance of social behavior, it is the hope of many scientists that understanding friendship's biology and evolution will bring us closer to knowing fully what it means to be human. "The message that's getting out is that if we spend our lives doing something"—connecting with other people—"there's probably some purpose it's serving for our species that's worth understanding," says Wheatley.[2] Neuroscientists, who are in the business of understanding human thought and behavior, are well positioned to do that work.

The social brain, as I've said, is a series of circuits, but let me expand on that. It includes all the components that are dedicated to processing interactions with other people. There is our sense of self, from the ability to recognize ourselves as distinct from others to our awareness of our own personality and of the autobiographical history we carry. There is our ability to take another's perspective (also known as mentalizing or theory of mind). There is our ability to regulate our emotions and behavior in response to other people or to fit societal norms. And there is our ability to detect whether we are being socially included or excluded, which amounts to a detection of threat.[3]

Each of these requires the integration of rapid, nearly automatic processes as well as more deliberative processes. We decode what we see and hear, touch, and taste almost instantly. Then we think about the information we've just received, weighing its value and importance, comparing it to our memories, anticipating what other people might do next or expect of us, and evaluating the social situation broadly.

Consider the following scenario. It is Saturday night. You step into a crowded party looking forward to some lighthearted fun. You spot a friend you haven't seen in some time, but you notice that he appears to be in a bad mood. He looks upset and is talking intensely to someone you don't know. You look around the room and consider your options: Approach your friend and try to find out what's wrong; head to another corner where some other guests you recognize but don't know well are laughing heartily; or, alternatively, introduce yourself to the woman you notice standing by the food looking lonely. The laughing group looks like the most fun, but what if they don't remember you? And will your friend be hurt if you don't come right over? Or maybe he will be annoyed if you interrupt his conversation? Talking to the lonely woman could be hard work, but also a kind gesture. Who knows? She might be interesting.

As in so many social situations, the best decision is complicated. There are plenty of unknowns, such as the mental state of your friend and everyone else at the party, and any reactions they might have to your decisions. Your brain was working hard on the problem from the moment you arrived. First you recognized people in the room and remembered past interactions with them (medial temporal lobes, fusiform gyrus). Then your mentalizing network kicked in, allowing you to make an educated guess as to what other people were thinking (superior temporal sulcus [STS], temporoparietal junction [TPJ], anterior cingulate cortex, medial prefrontal cortex). You considered what you wanted out of your evening, valuing your own preferences (ventromedial prefrontal cortex, orbitofrontal cortex, ventral striatum). You had to account for the value of your relationship with your friend (ditto). And you had to make some decisions and then assess the outcome (dopaminergic midbrain, striatum, anterior cingular cortex [ACC], and dorsolateral prefrontal cortex).[4]

Those parentheticals may have been a blur of unfamiliar names, but the fact that I can annotate this paragraph like that is the result of the last ten to fifteen years of neuroscientific effort. Researchers have been parsing the cognitive underpinnings of these capacities as precisely as my eighth-grade English teacher diagrammed sentences.

Neuroscientists have also been exploring what closeness looks like in the brain. Edith Wharton once said: "There is one friend in the life of each of us who seems not a separate person, however dear and beloved, but an expansion, an interpretation, of one's self, the very meaning of one's soul."[5]

In the 1990s cognitive scientists began to think there was something more than poetry to this idea. Art Aron and his colleagues, the same psychologist who developed the thirty-six questions that lead to love, hit upon a powerful way to capture that perceived merger of self and other. Instead of asking people to use words to describe the closeness of their relationships, they presented them with a set of seven Venn diagrams—pairs of circles representing the self and the other, displayed with increasing degrees of overlap from entirely separate to almost one. The choices people made proved a surprisingly effective measure of intimacy for romantic partners, for family, and for friends.[6]

Does this mean that in our mental processes, we think of our intimates as we think of ourselves? Perhaps. It could explain why we help close friends. Doing so feels like helping ourselves. Other researchers have found that the more we conflate self and other, the happier we are when a friend succeeds rather than feeling jealous.[7] It is also known that our moods swing with the emotions of those to whom we are close. If my husband or one of my children is having a bad day, chances are that I am, too. I felt the pain of my friend Stephanie's divorce keenly. Sometimes we feel so close to other people that we mistakenly remember things that happened to them as having happened to us. One study asked students to rate themselves on personality traits, and then to rate their best friend, closest parent, and a celebrity. Asked later to recall how they had rated themselves, the students tended to mistakenly recall the traits they used to describe their best friends.[8]

It wasn't a leap for neuroscientists to begin testing such merging of self and other in brain scanning experiments. This idea has been most extensively explored by neuroscientists investigating empathy. That began in 2004, when Tania Singer and colleagues at the Max Planck Institute for Human Cognitive and Brain Sciences in Leipzig, Germany, published a groundbreaking paper in *Science* that compared, for the first time, brain activity in a person experiencing pain and then observing a loved one experiencing pain.[9] Sixteen women underwent functional magnetic resonance imaging while their male partners sat nearby. Varied levels of painful stimulation were administered via electrode to one or the other partner. A signal alerted the women when their partners were feeling pain. Some areas of the women's brains were activated only upon receiving pain themselves, but others—most notably parts of the anterior insula and the anterior cingulate cortex—lit up no matter who was hurting. Empathy activated the affective, or emotional, parts of the pain network, but not the physical sensation of pain. That study and the many imaging studies that followed indicate that our core ability to empathize begins with the way the brain represents our own internal states and evolved to include our perception of what others are feeling.[10]

The overlap between self and other shows up in the brain in other ways as well. In one study, when people made judgments about their best friends, they activated areas of the brain also known to be involved in thinking about the self, whereas thinking about people they were less close to did not generate the same activity. In 2012, James Coan and his colleagues at the University of Virginia tested self–other overlap when faced with threat. Twenty-two pairs of friends came into the lab and were subjected to shocks on the ankle while holding the hands of their friends or of strangers. The scientists found that the way the brain responded to threat—how similar the activation was when the threat was to the self or to another—depended on familiarity and corroborated the psychological suggestions of an overlap. It suggested, they concluded, that "familiarity involves the inclusion of the other into the self—that from the perspective of the brain, our friends and loved ones are indeed part of who we are."[11]

Parkinson and Wheatley have tested whether the metaphors we use to describe closeness and distance between friends are reflected in brain processing. "Does our brain literally map out friendship?" they asked. Yes, it does. Seeing an object nearby triggers the same brain patterns as seeing a close friend. The response to objects that are farther away looks more like the response to seeing an acquaintance.

∽

In addition to overseeing the work on Cayo Santiago, Michael Platt runs a busy neuroscience laboratory at the University of Pennsylvania exploring how our brains shape how we think and how our lives shape our biology. "If you're a critter that's going to evolve complex social behavior, you're going to live a long time and your life is going to depend on others, especially being astute socially, a good mind reader and a good friend-maker," he says. "That gets deeply built into the brain."

The lab encompasses a suite of rooms lining a hallway in a wing near Penn's Medical Center. A dozen or so rhesus macaques are housed in the rooms on the right. For entertainment, they get to listen to the radio. As a treat, they get an occasional movie (David Attenborough's *Planet Earth* is a favorite). The spaces on the left were built expressly for playing games. They are soundproofed, full of screens and joysticks, and tubes for delivering juice directly into the mouth without having to move. It's the monkey version of teenage-boy heaven. While the monkeys play, the humans keep watch from adjacent control rooms full of expensive computer equipment—eye trackers, video feeds, and high-resolution devices that record neural activity from electrodes placed in the brains of the monkeys (please note that the brain does not feel any pain because there are no pain receptors in the brain itself). The technology allows the lab group to follow the activity of individual neurons and see the neurological trees amid the forest of behavior.

Like Capitanio at Davis, Platt believes that these animals have something important to tell us about ourselves. He describes his lab as operating at a scientific "sweet spot" between putting humans into MRI machines, which is useful but limited by what people can do while lying still and alone, and the dazzling molecular genetic work being done in mice, rats, and fruit flies. The latter allows scientists to selectively study particular neurons, but the work is hampered on the behavioral side by the very real constraints on the way mice and fruit flies socialize. "[Such work] doesn't really get into altruism, charity, love, the joy of being with other people, friendship," Platt says. "That's a big chasm." He believes he can get far closer to these concepts.

"Can we really study all those things in monkeys?" I ask.

"You can dig for the roots of some of those processes because the biology and much of the behavior seems to be shared," he says. "Our nature is fundamentally a primate nature. Our need and capacity to make friends with other individuals emerges from an ancestral adaptation." His work grows from the evolutionary work of his mentors Seyfarth and Cheney and others demonstrating that nonhuman primates have friends and allies, and that to have such relationships they need to know how to bond and cooperate. Summing up the results that he is building on, Platt ticks off the critical points one by one: "Those friends and allies buffer stress. The better their connections, the longer they live, the more babies they have. They're more successful. These skills and temperament that allow an individual to make these kinds of connections are biological so they're heritable."[12] What remains is to better understand what's happening in the brain.

At the lab, I watch as postdoctoral fellow Yaoguang Jiang prepares a macaque named Leakey, named for the famed paleoanthropologist, to play a video game that is a cross between soccer and hockey. The object of the game is to "kick" a ball through a goal using a joystick, but the ball ricochets off the walls like a puck. Sometimes Leakey and his peers play the kicker and sometimes they are the goalie. When they score, they get a squirt of juice. In another room, a macaque named Cajal, named for Santiago Ramón y Cajal, a pioneering neuroscientist who won the 1906 Nobel Prize (see the naming theme?), is playing a touch screen game set up on an iPad attached to his cage. Cajal has to touch a target while researcher Naz Belkaya measures his reaction time, which should vary according to the social value of the distractors that also appear on the screen (faces and genitalia of other monkeys). Cajal is rewarded with a pellet-sized treat every time he successfully touches the target. Yet another game, designed by postdoc Wei Song Ong, requires monkeys to drive simulated cars in a game of chicken. "Think James Dean in *Rebel without a Cause*," says Platt. In Ong's version, the monkeys don't drive toward a cliff as in the 1950s Hollywood classic. They drive straight toward each other (risking a collision) or veer sideways (avoiding a collision).[13] The question is, will they cooperate?

These games-monkeys-play are designed to plumb the complexities of the social brain. All are set up to deliver rewards. A simple dictator game trained monkeys to choose different shapes in order to deliver different sets of rewards—to the monkey itself, to another monkey, to no one, or to some combination of the above. On average, the monkeys proved to be very prosocial, significantly preferring to reward another monkey over no one. "There is something about rewarding another monkey that is motivating, that can add to the reward that monkey would otherwise experience for himself," Platt says. And the monkeys are more giving and less competitive with friends. The social motivation underlying this behavior was even clearer when Platt and his colleagues recorded from individual neurons in various parts of the brain known to be involved in social behavior, such as the superior temporal sulcus, anterior cingulate gyrus, and amygdala. "What's going on is that neurons in the STS are decoding the social context: who's around, what's at stake for whom and how's this affecting the other individuals," Platt explains. "And the neurons in the ACCg are decoding what the experience of another individual might be. And maybe the amygdala is mobilizing emotional responses based on those outcomes."[14]

It also very much matters who a monkey is playing against. Sometimes it is a computer, sometimes it is another monkey, and sometimes it's a "decoy" monkey that is in the room and looks to be playing but whose joystick is not actually connected to the game. Monkeys' brains respond differently to each condition. During my visit, Leakey played against a computer. That usually leads to very stereotypical play—simple strategies of getting to the ball and kicking it straight. But when monkeys play each other, their strategies become more complex. "They seem to try to fake the other monkey out, to fake left or right," Platt says. When presented with a decoy player, the monkeys quickly pick up on the disconnect between the decoy's facial expressions and other social cues and what's happening in the game. "It's social," says Platt. "Something happens between them—some communication and they look at the same thing."

Humans working together rely on similar social cues. It starts by looking. Primates are very visual creatures. "We communicate visually, and our visual attention is critical to how we develop," Platt says. "One of the first things that seems to fall apart in autism is attention to others, then joint attention follows from that, and language."

Eye contact is a favorite topic of Platt's and it's a bit of a running joke among the members of his lab, several of whom are standing by while we talk.

"Being a student of someone who's an expert in social gaze can sometimes make it very awkward when you're in lab meetings," says postdoctoral fellow Geoff Adams. "Where do I put my eyes?"

"I'm very conscious of everybody's gaze," Platt admits, turning his blue eyes on each of us in turn for an exaggerated few seconds.

"We know," says Adams. The room erupts in laughter. "We're conscious of you being conscious of it."

The monkeys are, too. When they play competitive games with joysticks, they watch each other intensely. "As soon as they start playing, they're watching the other monkey all the time. They watch his eyes. They watch his hands. They watch him drink juice," Platt says.

In Jiang's game, the animals consistently glance to the spot where they intend to kick the ball several seconds before they act (that's how the scientists can tell when they're faking). To each game, monkeys also bring their knowledge of relative rank and their history with this particular animal. The friendlier two monkeys are, the more attentive they are to each other, and more motivated they are to share.

In a particularly exciting result, Ong and Platt found a population of neurons, again in the STS, that responds significantly more when the monkeys achieve the same size reward but do so by cooperating. The chicken game is set up so that it is possible to tweak the strength of the signals the monkeys receive from one another's actions on the screen. "There is a lot of opportunity for the monkeys to coordinate their behavior using these intention signals," says Platt. The weaker and more random the signal, the more crashing the monkeys do. "That is consistent with some evolutionary signaling theory that says the reason we have signals and displays is to avoid outright conflict." They also found that cooperation activated cells in a brain region linked to strategic thinking rather than, say, empathy.

While monkeys don't make all the same social decisions as you or I would exactly, they do share many of the same brain mechanisms. Platt's team is taking advantage of the similarities to uncover specializations in the social brain. Just as babies prefer to look at faces, rhesus macaques will forgo juice for the pleasure of looking at a picture of another monkey, and the scientists can see that such a decision stimulates reward areas in the brain. Furthermore, different neurons fire depending on who gets a reward in a game—the monkey itself, another monkey, or no one.

Lesions in parts of this social brain can lead to social deficits. Understanding those social deficits is one of Platt's primary goals. "The study of connection is interesting in its own right," Platt says. "But it's also terrifically important as we begin to think about ways of helping people who have difficulty making these connections." The ultimate goal of one line of research on Cayo Santiago is to use the genomic and social network analyses to identify animals beyond the boundaries of "normal" social behavior that also have the related genetic variants. Then Platt hopes to bring monkeys with the same set of alleles into the lab at Penn and examine the fine-tuning in their brains. "Most of us in the field think there's a set circuitry and then you have a volume knob. The big question is what turns it up or down."

⤳

One likely answer is the brain's own biochemistry.

The brain runs on neurotransmitters, including the group sometimes collectively referred to as the happiness hormones: oxytocin, endorphins, dopamine, and serotonin. Technically, a neurotransmitter is released by a nerve cell and a hormone by an endocrine gland, but they are essentially the same. Neurotransmitters are brain hormones. They've been dubbed "the wireless network" for the body by science writer Randi Hutter Epstein.[15] They have the capacity to turn the volume knob that Platt described up and down, but they also have the capacity to break it.

The most famous neurotransmitter these days is oxytocin, which has gotten substantial positive press as the love hormone and the moral molecule. Oxytocin owes its burst of celebrity to the lowly prairie vole, a species known for its strong social nature. For decades, oxytocin was known only as a factor in childbirth and breast-feeding (surges in oxytocin bring on both labor and lactation), and as an important element in the bonding of mothers and infants. But in the early 2000s, neuroscientist Larry Young of Emory University and his colleagues discovered that oxytocin does more than that. When they injected it into the brains of voles and other species, it "worked like a mythical love potion, creating an instant and powerful monogamous attachment," writes Paul Zak, a leading oxytocin researcher.[16] In fact, oxytocin served to strengthen all forms of attachment, not just bonding with a mate. Among other things, Young showed that the administration of oxytocin enhanced social recognition in mice, and if you manipulate a mouse's genes so that it no longer produces oxytocin, social recognition falls apart.[17]

Platt's team figured out how to get monkeys to inhale oxytocin by repurposing a pediatric nebulizer. Then they gave monkeys a reward/donation task and the monkeys became more giving, more empathetic. They showed more vicarious reward. "They also pay more attention to other monkeys, they look at them longer, they look them in the eyes," Platt says. "The oxytocin seems to encourage the kinds of social interaction and attention to others that go awry in kids with autism." Their preliminary data suggest that what the oxytocin is doing is making the communication in brain circuitry clearer. "It's higher fidelity in the sense that we have more signal and less noise."[18] (Multiple clinical trials are under way exploring the effectiveness and safety of inhaling oxytocin as a therapy for autism. Hopes have been high, but they have so far yielded mixed results.[19])

For all the positive press that oxytocin gets, it comes with terms and conditions. Cathy Crockford and Roman Wittig of the Max Planck Institute for Evolutionary Anthropology in Liepzig, who like Platt learned their craft in the lab of Seyfarth and Cheney, discovered recently that surges in oxytocin depend on the individuals involved. They studied pairs of grooming chimpanzees. If any old troop mate was doing the grooming, hormone levels didn't change much. If it was an individual with whom the recipient had a close bond—including but not limited to kin—oxytocin levels rose considerably.[20] What mattered most, in other words, was whether the chimpanzees were friends. That is an important result. "This is starting to say that there's something about interacting with individuals that you perceive as close friends that's physiologically very rewarding," Seyfarth says.

It's also possible to have too much oxytocin flowing through the brain. A study published in 2016 in *Science* pinpointed the processing of oxytocin in the brains of prairie voles. The work got under way when James Burkett, a neuroscientist at Emory University, noticed that the voles seemed to be consoling each other when stressed. He tested the observation more explicitly in an experiment. Pairs of animals were caged together for a few weeks, then the female was removed briefly. She either was simply kept separate for a few minutes or was given a mild foot shock, a form of fear conditioning that generates stress. When the animals were reunited in the cage, Burkett's team observed their natural social interaction. If the female had not been stressed, neither animal seemed particularly anxious. But when she had been shocked, the male quickly began grooming her at high levels—behavior interpreted as consoling because the pair did not engage in it otherwise. The animal left behind showed a physiological response that mimicked that of the animal that had been taken away. Furthermore, the intensity of the consoling response varied from animal to animal.[21]

When Burkett looked at the brains of the voles, he discovered that their responses correlated positively to levels of oxytocin signaling and negatively to the density of oxytocin receptors in the same part of the brain—the anterior cingulate cortex—that Tania Singer had identified in humans who felt pain empathy for others. There appears to be a sweet spot for oxytocin levels. It is possible for the brain to be overwhelmed—in effect, swamped—by oxytocin. As a result, animals with high oxytocin receptor density consoled for shorter periods of time and vice versa. "The greater the personal distress an individual feels," says Burkett, "the less likely that individual is to help."[22] Burkett's study provides a possible explanation for one of the negative aspects of empathy. When the emotions being experienced are stressful or painful, empathy is painful. "If I empathize with everyone who is in a worse state than I am, I might be motivated to donate 95% of my income to charity," says Stanford's Jamil Zaki. "Rather than be put in a moral double bind between guilt and poverty, I might just choose not to think about people who are less fortunate than myself."[23] In certain professions, such as medicine and law enforcement where exposure to human suffering can be constant, too much personal distress gets in the way of doing the job. Physicians, for example, suffer from excessive burnout and are at higher risk than others for death by suicide.

In our good friendships, however, the levels of oxytocin, endorphins, and the other happiness hormones seem to be calibrated just about right. They are triggered by laughing, by singing together, and by storytelling.[24] They make us feel good and make us want to come back for more.

~

Let's return to Brad, who has been waiting patiently in the scanner. He will be in there for about forty-five minutes during which the researchers will run eleven sequences of functional magnetic resonance imaging. Some are intended to capture the anatomical details of Brad's brain. I can see a hint of this on a computer screen in the control room. First, there are three low-resolution images and then a set of higher-resolution pictures making everything crisper and brighter. These are slices of Brad's brain seen from the side, from the front, and from the top—the sagittal, coronal, and axial views. In all my years of writing about the brain, this is the first time I've seen such images live, as it were, rather than on a page or in a photograph. There is Brad's brain stem and his cerebellum and his corpus callosum, which links the two brain hemispheres, standing out in white. I can see the bumps and folds of his frontal cortex. I'm enthralled. The overall shape of Brad's head clearly suggests the same long, narrow face that I saw on the big screen earlier as he lay down. But now I can see the inner workings . . . working.

After about eight minutes in the scanner the anatomical scanning is complete, and Brad begins to watch videos. There are six sets of clips in all, unconnected videos culled from every corner of the internet. As a whole, the exercise is like channel surfing, says Carolyn Parkinson, a shy Canadian just past thirty off to an award-studded start on her career.[25] The first clip is from an interview with a Canadian (naturally) astronaut named Chris Hadfield who describes the Earth as seen from space. There are head shots of Hadfield and images of Earth from space. Some are exquisitely beautiful, but others reveal the pollution hanging over big cities like Mexico City and Beijing. "It's like a big grey smear on the face of the Earth," Hadfield says. Next is a snippet from a music video. Then part of a dryly comic Australian mockumentary about an unimpressive man who nominated himself to be Australian of the Year. There is footage of volunteers and staff at a sanctuary for sloths. "He likes to be slow-danced," says a woman gently rocking back and forth while holding a sloth identified as Sid. There is a video review of Google Glass. There is a skit from the *Tonight Show* featuring Jimmy Fallon, Seth Rogen, and Zac Efron dressed as teenage girls and taking selfies. When I look it up online later, the review I find says "We don't know if this is funny or downright disturbing. Proceed with caution."

Whether you think it is funny or downright disturbing will depend on your sensibilities, which is exactly why Parkinson chose it. All of the videos are likely to be captivating but also to engender strong responses. When her research assistant checks in with Brad after the Australian mockumentary, we can hear that he is still laughing. And therefore, if I had to guess, I'd say that Brad was less interested in the sloth sanctuary. Here comes the important part: What Parkinson and Thalia Wheatley have already discovered is that, several months from now, once classes have started and Brad has made some friends, it is likely that the brains of those friends will respond to the videos much as Brad's did.

They reported that discovery in *Nature Communications* early in 2018.[26] The paper was Parkinson's dissertation. To do the research, she and Wheatley joined forces with Adam Kleinbaum, a professor at Dartmouth's Tuck School of Business who studies social networks. After creating an analysis of the social networks of a large group of graduate students, they put forty-two members of the group into the scanner. While people watched the same video clips Brad is now looking at, the scientists divided their brains into eighty different anatomical regions of interest, such as the amygdala, and tracked how the response rose and fell over time for each region. Parkinson compared the resulting set of time series from corresponding brain regions for each pair of people to look at how similar their responses were within each brain region. Because each person saw the same clips in the same order, any differences in their brain's responses were theorized to be the result of differences in personality or perspective.

As we know, similarity has been recognized as a hallmark of friendship since at least the Ancient Greeks and presumably before that. When scientists have tested the assumption over the years, they have found that, indeed, people are more likely to befriend others of the same gender, age, ethnicity, and so on. "It's such a pervasive tendency across so many eras and locations and societies to surround yourself with similar others," says Parkinson. Perhaps that's why, although my friends are fairly diverse, they include a lot of white, middle-aged mothers with creative jobs. And why, when I met Wheatley—who checks all those boxes—and we started talking, I quickly felt that we could be friends. Similarity explains the friendship between Gail Caldwell and Caroline Knapp, beautifully captured in Caldwell's memoir *Let's Take the Long Way Home*. They were both single, journalists, recovered alcoholics, and passionate dog owners. A distinct lack of similarity between friends foretells enough conflict to be sitcom gold. (See *The Odd Couple*.)

As it turns out, that similarity extends all the way to the firing pattern of your neurons. Parkinson, Wheatley, and Kleinbaum could predict which participants were good friends and which were not by matching up the way their brains perceived and responded to the world around them. That correspondence didn't just show up in regions known to process social information. "It was all over the brain, sensory regions, memory, language," says Wheatley. "You couldn't say here's the social brain network that is responsible for friendship. It was that friends' brains are remarkably similar across huge swaths of areas." The correlation was strong in the areas that govern what you look at or listen to. "Your friends might be paying attention to the same parts of the [video] or deploying their attention in similar ways," says Parkinson. That surprised me. Why should the pattern in a place like the auditory or visual cortex be more similar? Don't we all hear the world the same way? Apparently not. "It's getting at some sort of experiential richness," says Wheatley. "Friends are literally seeing and hearing the world more similarly than people who are friends of friends, and friends of friends of friends. It's coming down to the level of how you're processing sights and sounds. Given a music video, if you and I find some part of the melody or some part of the visuals particularly engaging, then we're going to tune our eyes and ears to those parts. That's remarkable."

Pondering this, I thought about my drive to Dartmouth with my middle son, Matthew. We were combining my reporting trip with college visits and had been listening to the Ink Spots, a 1940s quartet, as we drove into town. The Ink Spots were one of my father's favorites so now they are a nostalgic favorite of mine. Matthew, who plays jazz saxophone and has eclectic musical tastes that encompass everything from Frank Ocean to Frank Sinatra, is the only one of my three sons who doesn't just put up with my old-fashioned taste, but truly seems to enjoy the baritone crooning of "Into Each Life Some Rain Must Fall." Maybe we hear it the same way?

"If you put the Ink Spots on in the car, you and your middle son would have the same sort of experience," Wheatley agrees. "Even though it's the same stimulus, another one of your sons would have a completely different experience."

Of course, this doesn't mean that I don't love my other two sons as much as I love Matthew. Or that I can't be friends with people who don't share my musical tastes. My dear friend Moira is passionate about Bruce Springsteen, which I am not, although we do have a similar sense of humor. But it is true that listening to the Ink Spots, as well as Édith Piaf and Patsy Cline, is a special part of my bond with Matty.

"There are advantages to surrounding yourself with similar others," Parkinson says. "Potentially, people [who are alike] might share similar goals and assumptions and experiences and that could help foster cohesion, empathy and collective action. It's interesting to think about."

Another striking result from that study was that the farther away from each other in the social network two people were, the less similar their brain responses became. But that linear progression was only true up to three degrees of separation, a pattern that reinforced Christakis and Fowler's rule of three degrees of influence. "It breaks down once you don't really know people," Wheatley says. "If you got to know them, you might actually pull them close to you. Conversely, what we do know is that you actually do know people who are friends of friends and people who are friends of friends of friends well enough to hold them at arm's length. You know them well enough to know that your brains don't work similarly enough to be any closer."

Parkinson calls that first study "proof of concept"—they showed you could put people in a scanner one at a time and draw intriguing conclusions about the connections between them. But that work captured only a snapshot of a moment in time. The results begged the larger question: What comes first? Are friends with similar brain patterns drawn to each other? Or do their neural responses change because they're friends and spend time together? So far, we don't know.

This is where Brad comes in. He is participating in a new longitudinal study Parkinson and Wheatley hope will shed light on this question of direction. This time around, the scientists recruited a second group of graduate students and are scanning some of their brains *before* classes begin. Then they will follow the group for the year to see who befriends whom and they will repeat the brain scans to see whether they can pinpoint how an individual's responses change after exposure to new friends. Did they start out similar—both finding the Australian comedy hilarious? Or did one person's impassioned arguments on behalf of the environment change how someone else looked at those photographs of pollution seen from space? It seems very likely that the effect must go both ways. We may well be drawn to those who process the world in much the same manner as we do, but our minds might also be changed, literally, by our subsequent association with those people.

Alas, I can't tell you the answer here. The results of the study in which Brad is participating will take time to gather and analyze—more time than my book deadline allows—though the study itself tells us where the science is going. However, another recent study from Wheatley's lab provided some related clues. In this one, participants were put in the fMRI scanner to watch snippets of a movie with the sound off. The clips were chosen to be ambiguous. "You can't really tell what the relationships are of the characters on the screen," says Wheatley. After watching parts of the movie, participants came together in groups of five to discuss what they'd just seen. They were asked to come to consensus. Were the characters brothers? What exactly had happened? "You spend an hour or two hashing all these scenes out with each other, figuring out what you agree is the right interpretation," Wheatley says. Each group came up with notably different interpretations. Then everyone went back in the scanner to watch the same clips again as well as some clips from farther along in the movie. Each group's agreement was visible in the brain. "Everybody's brain looks very idiosyncratic at the beginning. And then, after you've had a conversation, you come into a kind of neural alignment," says Wheatley. "That group sees the movie the second time, and they see those movie clips as a single brain, in the same way. People are becoming aligned through a shared experience, due to conversation."[27]

Perhaps it's not surprising then that when graduate student Emma Templeton came to Wheatley and said she wanted to study what made conversation fun, Wheatley was fascinated by the idea. They are video recording hundreds of one-on-one conversations between strangers and are transcribing every word and coding the videos. Like the brain imaging work, this study won't be complete before I'm done with my reporting, but they did arrive at a preliminary finding: conversations rated more fun by the speakers are marked by rapid turn-taking. "We think it is an honest signal of engagement and shared understanding," Wheatley says.[28]

———

The kind of brain synchrony that Parkinson and Wheatley are uncovering really is the state of the art in neuroscience these days. It's been found in other ways as well. When people attend a concert together, or take part in an engaging class, scientists have found that their brain wave patterns—those that can be measured by EEG—become synchronized.[29] Neuroscientist Uri Hasson of Princeton University has been a leader in studies of how two brains interact. In one pioneering study, he put individuals in a brain scanner and asked them to tell their life stories. Then he put speakers in the scanner and played them recordings of the stories. He found evidence of the two brains coupling—aligning their activity during the listening, which he considers communication. The more aligned the coupling, the greater listeners rated their comprehension of the story.[30]

In order to watch two brains interacting simultaneously, Joy Hirsch, a neuroscientist at Yale, has turned to functional near-infrared spectroscopy (fNIRS), the same technique that Sarah Lloyd-Fox and Mark Johnson were employing to investigate the brains of babies. Hirsch got a Japanese company to build her an fNIRS setup that completely covers two adult heads at once. Using it to study eye contact, she and her colleagues discovered that when two people look at each other, it triggers different neural activity than looking at a picture. And she found something else, too. "If you are looking at a real face, the really neat thing that happens is that elements of language system are activated," she says. "I think it means direct eye contact with another person is a call to action, a call to speak, and readiness for engagement."[31]

But of the handful of scientists pursuing multibrain interactions, Thalia Wheatley is one of the few focused on friendship. "We spend our lives having conversation with each other and forging these bonds," she says. "[Yet] we have very little understanding of how it is people actually connect. We know almost nothing about how minds couple."

A good conversation, says Wheatley, means creating something new together. "It's a walk in the woods and you don't know where you're ending up. You're creating new ideas together and experiences you couldn't have gotten to alone." Collaboration in science is a perfect example, she says, because it is done collectively. "We all get our students together. We brainstorm. That's the way we think more intelligently and creatively, with other minds."

But the actual science, the research, does not yet capture that process. Echoing Platt's complaints about the limitations of fMRI, Wheatley says, "We devise these experiments that are stuffing one person, supine, in a noisy tube and showing them images. We study the brain as if it works as an isolated unit. It's sort of a brain in jar sort of approach." As she points out, the research that I observed is still actually only testing one person lying in the scanner. In Hasson's earlier work, that person is telling a story to which someone else will listen. "Then you match them up over time and you see, okay what the speaker's brain was doing looks like what the listener's brain was doing," says Wheatley. "But those people could have existed 100 years apart from each other. They could never have met. They're not creating . . . it's not a conversation." If she has her way, she will soon capture the next level of interaction. "How do minds interact in real time to create this shared understanding and shared experience with each other?"

She is joining forces with Hasson to pursue a technique called hyperscanning. It allows two different people in two separate fMRI machines to be scanned simultaneously while carrying on a conversation. The technological difficulties have been considerable and the math daunting. But Wheatley and her colleagues are closing in on making it work to study friendship. They have a working framework for connecting scanners at different universities over the internet. During my visit, Wheatley conducted a video chat with Adam Boncz of Central European University in Hungary who recently spent a year in Wheatley's lab and is spearheading their hyperscanning effort. They have already succeeded in doing a few pilot hyperscans using the scanner down in the basement and one at Harvard University. Wheatley has been a subject in these preliminary tests. She lay down in the fMRI at Harvard while a graduate student lay in the scanner at Dartmouth. They endeavored to complete a story, with one starting off and then the other picking up the story in turn. "It worked beautifully," she says.

And this is Thalia Wheatley's dream. "What I want to know is when our brains are dancing together, what does that look like? And why do some people dance together better than others? Why do we click with some people more than others?" She and Boncz say their rough idea is that if they can capture the states that the brain revisits again and again during an interaction, they might be able to see the dynamics of how two brains get closer and further away from each other. They might just see a friendship forming in real time.

TEN

The Good Life, Revealed

One day when she was in her late sixties and at home alone, Paula Dutton's heart started racing and she began to have trouble breathing. Feeling faint, she began to have pain in her chest. She was terrified.

"I really felt like I might die," she said later.

She managed to call for an ambulance. When the paramedics arrived at her house in South Los Angeles, however, they realized that Dutton was having a panic attack, not a heart attack. They calmed her down and she didn't have to go to the hospital.

Dutton wasn't dying, but she knew she was in trouble all the same.

After decades working for the phone company, Dutton had retired in 2011. "At [work], there were always people around," she says. "You always had that camaraderie with your fellow employees." Now she had no colleagues and she had no family nearby either. A ten-year marriage had broken up. She had never had children. She lost first one parent, then the other. Her remaining family was back on the East Coast in Philadelphia, where she had grown up.

"I just suddenly realized that I was all alone and had no one around me and no one I could turn to," she said. "I had a lot of pity parties, I can tell you, and with all kinds of anxiety and depression, and I worked myself into a fever pitch in my loneliness. I had gotten to where with the anxiety and the bad feelings, I thought: Is being so lonely making me sick?"

The episode prompted Dutton to make some changes. She started by joining a church near her home, which gave her a sense of community. Then in 2015, she ushered in two more changes. She met a new man, whom she is still seeing. And she joined a program called Generation Xchange, which brings her into the 74th Street School, an elementary school just a few blocks from her home, four or five days a week.[1]

As I follow Dutton around Theresa Brissett's first-grade classroom at 74th Street in the fall of 2018, she is no longer scared or lonely.

"Miss Paula, Miss Paula, look what I made," calls out one girl, who has constructed a tall building from plastic blocks. It is her physical model of the vocabulary word displayed on an index card in front of her: "tower."

Dutton makes her way around the table to admire the tower. With her cell phone, she snaps photographs of the tower and of a house made by the boy in the next seat. She tucks the phone back into her pocket and turns to me. "When I changed phones five months ago, I had 1,100 photographs from school."

Small and trim, with tightly cropped hair, Dutton is casually but fashionably dressed in jeans and red flats. She is seventy-three when we meet, but the only sign of her age, other than the sprinkling of gray in her hair, is some stiffness in her knees. That doesn't stop her. At the next table, she kneels down next to a girl in pigtails who is reading. Quietly intent, Dutton asks about the book and the girl's face lights up with a wide grin as they talk. Dutton grins back. Another boy is building a rocket out of Play-Doh and toothpicks. "Is that going to go up to the moon?" she asks him. Out comes the cell phone camera.

"They all feel like their grandma is present," Brissett tells me, and she can relate. Brissett and Dutton have grown so close over the two years they've been working together that Dutton has become something of a substitute grandmother for Brissett's own eight-year-old son.

"I'm originally from the Caribbean," Brissett says. "My Mom is so far away. My mother-in-law is so far away. Miss Paula is like my mother here in California."

Generation Xchange changed Dutton's life. That's just what it was designed to do, in ways that are both obvious and subtle. The intergenerational program is the brainchild of UCLA epidemiologist Teresa Seeman. It is an educational nonprofit wrapped in a community health initiative with a loneliness intervention program beating quietly but steadily at its heart. To create Gen X, as it's known, Seeman's employer, UCLA's Division of Geriatrics at the David Geffen School of Medicine, joined forces with Los Angeles's Unified School District. They bring older adults from the local community into underresourced elementary schools throughout South Los Angeles. Like the neighborhoods they are in, the populations of the four schools served so far are predominantly black and Hispanic. An average 40 percent of their students are from foster care homes. While some schools have gifted programs attached, many of the students struggle and need extra attention.

After training, each Gen Xer, as the participants call themselves, is assigned to a specific classroom—anywhere from kindergarten through third grade—for the entire school year. They spend a minimum of ten hours a week with the children, though nearly everyone puts in more time. As an educational program, Gen X aims to improve children's academic skills, especially reading, and to address behavioral issues. As a health initiative, the program is intended to get the older adults out and moving. It targets cholesterol and blood pressure, weight loss, and increased mobility. As an intervention for loneliness, it is designed to foster connection and build friendships. It has succeeded on every front and even in ways (spoiler alert: it has to do with gene expression) that Seeman didn't imagine when she started it.[2]

As we age, the effects of the life we have lived—including our relationships—inevitably show up in the body. Some are cumulative, some are short term. Remember that John Cacioppo found worrying signs in the vascular systems of lonely college students but in lonely older adults those early problems had had time to fester and do harm, resulting in high blood pressure.

Teresa Seeman has spent her career thinking about social relationships and the later stages of life. In the 1980s, fresh out of graduate school, one of the first things she did was revisit the data from the Alameda County study. It was in Alameda that Lisa Berkman and Leonard Syme showed, for the first time, that social relationships were associated with mortality. Seeman was greatly influenced by their work. She arrived at Berkeley to begin her doctorate in epidemiology just as Berkman was leaving with hers in hand, but the two would go on to be regular collaborators. They were the ones who suggested that maybe what social isolation was doing was speeding up the aging process.[3]

While Berkman's early work showed that relationships mattered generally, Seeman had a lot of questions about the specifics. Seeman also noticed that no one had looked at how the effects of relationships might change as people grew older. The seminal 1979 paper by Berkman and Syme included adults who had been thirty to sixty-nine years of age at the start of the study. What is true of people in their forties and fifties, however, does not have to be true for those who reach their seventies and eighties. Seeman wondered whether the increased risk of mortality that came with social isolation continued into older age groups. And she wondered whether one type of social tie was more important than another. Did you have to be married? Were other relationships equally effective? She decided to analyze the data from Alameda County by age group and see whether she could work out what had the strongest effect on the oldest people in the group. She was especially worried about widows and widowers. "As people got older the probability that you would lose a spouse got higher," she says. "So, if marital status continued to be super important that would suggest there was a whole lot of people being exposed to increased risk and that would not be good."

The Alameda data included questions on four types of social ties. Three out of four of those questions had simple yes/no answers. Are you married? Yes or no. Do you belong to a religious organization? Yes or no. Do you belong to other kinds of community organizations? Yes or no. The fourth question, about close friends and relatives, had a little extra meat to it. It asked how many close friends and close relatives people had, and also how many of these people they saw at least once per month. Those who answered that they had five or fewer interactions per month with close friends and family were considered "isolated" (a group that amounted to 24 percent of the sample). Berkman's original 1979 paper had looked at mortality going out nine years. By 1987, Seeman was able to review seventeen years' worth of outcomes.

When Seeman did her reanalysis, she included everyone who had been anywhere from thirty-eight to ninety-four years old at the start of the study. She found that both age and the type of relationship did matter—quite a bit, in fact. For those under sixty, marital status had the most significant effect. In other words, being unmarried in midlife put people at greater risk of dying earlier than they otherwise might have been. But—and it was a big but—this did not turn out to be true for the oldest groups. For those who were over sixty, close ties with friends and relatives mattered more than having a spouse.[4] "For me, that was a real lightbulb that went on," Seeman says.

Three decades of subsequent work have only reiterated the importance of that early finding. "The most important thing about social relationships is just how critically important and valuable they are," Seeman says. "Earlier in life, being married—that relationship—is really key, but as you get older friendships become that much more important and whether or not you're married is relatively less important." This is good news. Even if you lose a spouse, friendships can sustain you. And you can keep making new friends throughout your life.

That is exactly what the women who volunteer at 74th Street School have found. To a person, participants in Generation Xchange say they have made new friends and are less lonely. In addition to the relationships

the volunteers build with children and teachers, they develop relationships with one another—the same easy camaraderie people like Dutton once had at work.

That camaraderie was on display on the October morning I visited.
A group of ten women sat around the small tables in an empty classroom. An eleventh volunteer was away with her class on a field trip,
but everyone else assembled for the weekly team meeting. (Although
staffed primarily with women, the program has male volunteers working in other schools and is looking for more.) As always, there were
snacks and drinks.

Linda Ricks is the lead volunteer for 74th Street and she runs the
meeting with a firm hand and a firmer sense of humor.

"This one is my problem child," she tells me, pointing to a woman
named Barbara Phillips.

They've been discussing a potential group lunch before a school
event.

"That means you're paying for lunch, right?" Ricks says to Phillips.

"I'm paying you no attention, that's what I'm paying." The group
erupts in laughter.

"See I told you she's my problem child."

"You done met your match," another woman observes.

Not likely. Ricks is seventy-four. A widow and now a great-
grandmother, she is something of an *über*-volunteer. When she retired
twenty years ago from her career as a clerk at the post office, a federal
agency, and finally a bank, she started as a reader for the Grandpar-
ent and Books program at her local library and spent years there. For
the last ten years, she has also worked as an activist in the black community, holding voter registration drives and informational meetings
about candidates and propositions during political campaigns. As soon
as she heard about Generation Xchange, she told D'Ann Morris, one of
Seeman's codirectors who handles recruiting, to count her in.

Ricks has been in the program for five years and is now in her third
year working with Michael Travers's kindergarten class at 74th Street.

On the day she met this year's crop of children, she says, the kids all yelled, "Mr. Travers, Mr. Travers, is she your grandmother?"

Ricks cuts her eyes at me. "Mr. Travers is white."

He is also a big fan of hers. "When she arrives it's like the Beatles on the Ed Sullivan show," Travers tells me. "They would mob her if I let them." Both Mr. Travers and Ms. Brissett have instituted a finger-wave-only policy for when Dutton and Ricks arrive each morning.[5]

Ricks is in the classroom Monday through Thursday mornings while the children rotate in groups through the reading stations. Her style couldn't be more different from Dutton's gentle attentiveness. Ricks is the demanding grandmother who doesn't put up with any nonsense. Her voice rings out regularly through the classroom. While Travers works with one small group in the opposite corner, Ricks oversees another group matching pictures to letter sounds. "What is that a picture of?" A lock. "Where does it go?" Columns headed J, K, and L are on each child's sheet. When kids successfully sort their sounds, they get a high five and a sticker from Ricks.

At the Gen X meeting, Ricks pulls out her agenda. There is new paperwork to discuss. Volunteers must record any children who move in or out of their classrooms. She notes who will help out at the upcoming Harvest Festival and Halloween Parade. Then the women chat about politics ("I'm not telling you who to vote for, but make sure you're registered," Ricks instructs). A possible school strike looms and the women discuss what they should do if it happens. They are not employees of the school district, but some feel strongly about supporting unions. Others worry about leaving the children.

Then they start to talk about how they came to be part of Generation Xchange in the first place. Although some were frequent volunteers like Ricks, there are echoes of Dutton's experience in many of their stories.

"I had been retired maybe fifteen years, and I was tired," Wynona Price says. "When my grandson would visit, I didn't want to do anything. He would say, 'Grandma, you need to get out and move around and go somewhere.'" On the day she started with Gen X, Price told Ricks she was probably going to have to have knee replacement surgery. But the increased activity that came with being part of the program has worked miracles. "Since I've been here, I physically feel so much better," Price says. "I haven't even thought about my knee replacement."

In turn, she has recruited her sister, Barbara Bass, who is the newest member of this group. "She thought I needed something to do, not just sit in the house and watch TV," Bass says with a laugh.

Everyone is retired. Some have lost spouses. There have been illnesses and infirmities. One woman had breast cancer. Menora Garner uses a cane. She makes her way from her house across the street. Her children and grandchildren attended this school and she wanted a reason to stay involved.

It is the relationships that keep them here. They tell tales of individual children. There is a boy who waits for Miss Bernice at the school door every day so he can carry her purse to the classroom. Another made Ricks laugh when he told her, "I know you are old. You've got old people hair."

"When they get something, they are so happy," Phillips says. "They come running and say, 'Miss Barbara, look what I did!'" Everyone has a similar tale of feeling loved.

And the love these women get from one another is as important as what they get from the children.

"These women are God's apology for my family!" Phillips announces. "They are my friends."

"You can't help who's in your family," someone else chimes in. Laughter fills the room again.

Even with all the fun and affection on display, the bond between Paula Dutton and Linda Ricks is special.

Unlike Ricks, Dutton wasn't at all sure that Generation Xchange would be for her. Three years ago, right around her seventieth birthday, her friend Bertha Wellington urged her to go to an orientation meeting. After the panic attack scare, Dutton knew she needed to do something more, but she was reluctant to do *this*. With no children of her own and little experience with young people, she couldn't imagine working with children in an elementary school classroom for hours every week. "This was the most frightening thing that I think I could do," she says.

As she arrived at the orientation, another woman reached the door at the same time. It was Ricks. They went in together, so it was natural that they would sit together. And Dutton could see that she was among peers. ("You can automatically identify with them.") At one point, Dutton and Ricks sneaked back outside together for a cigarette (shh, don't tell) and by the time one gave the other a ride home (it turned out they lived seven blocks apart even though they had never met), they were fast friends.

Ricks was full of encouragement.

"Listen," she told Dutton, "those little suckers will grab onto your heart before you know it."

Dutton decided to try it, but to stick with her new friend. "I decided I would go wherever Linda was," she says.

Since then, the two have always worked in the same schools. They sometimes go shopping or to lunch afterward. And despite her fears, Dutton proved to be a natural.

After the first day in the classroom, when Ricks asked Dutton whether she was staying, Dutton smiled and answered, "Girl, they need me."

That kind of success doesn't show up in the report that Seeman and her codirectors put together for funders even though the official results are impressive. In one school, the percentage of children who met reading standards by the end of the school year nearly doubled in kindergarten (from 46 to 78 percent) and first grade (from 38 to 77 percent) in the first two years that Gen X was in the school. And in second grade it rose from 61 to 78 percent. Teachers and principals reported improvements in attendance and ability to pay attention and manage social interactions. There were also fewer referrals for discipline. Meanwhile, in the older adults, the hoped-for health benefits have all come to pass. Blood pressure and cholesterol are lower. Mobility has improved (not just for Wynona Price, but for everyone). Many have lost weight. Ricks has dropped forty-five pounds. "My doctor is very happy," she says.

And there has been one other improvement that wasn't on the original list. At UCLA, Seeman knew of Steve Cole's work on gene expression in the immune system. She also knew that he had gone on to find that while loneliness and other forms of adversity increase susceptibility to inflammation and viral infection, social integration and feeling fulfilled seem to have the opposite effect, improving the immune system's ability to fight off those ills.[6] Given the highly social nature of Gen X, Seeman thought perhaps her volunteers might show such a change and Cole was happy to take a look.

All of the participants routinely give blood at the beginning, middle, and end of each school year. Beginning with the 2016–2017 year, Cole has run his genomic analysis on the samples to measure any change in their gene expression that corresponded to his earlier research. Sure enough, he found statistically significant improvement in the right direction: less inflammation and more protection against viruses. Being part of Generation Xchange has improved these older adults' health right down to the level of their genes.

I don't want to overstate these results. While Cole and Seeman hope to publish the findings, the sample is still very small. To make larger claims, they would have to compare the volunteers to a control group, which they have yet to do.

Still, it's encouraging. The friendships appear to be a critical piece of the program, but not because the Gen X volunteers are thrown together and told to interact but rather because they are engaged in a meaningful endeavor designed to bring them together with a shared goal. It's like taking vitamin D to make sure the body absorbs calcium—one works better with the other and the bones get stronger as a result.

That's why Seeman organized the program as she did, with time for group meetings and end-of-year parties. "We don't say we're trying to recruit you so you can make more friends," she says. "We say we're trying to recruit you because we all think that helping these kids is great. And by the way, we think it may have some health benefits for you. That's where we leave it." But asked whether her ulterior motive was to bring about friendships, she doesn't hesitate. "Oh yes."

⤻

The women of Generation Xchange neatly encapsulate the polarities of friendship and loneliness in the latter part of life. Paula Dutton's life is now full of joyous strong relationships with a wealth of new people— her gentleman friend, her fellow church members with whom she was about to travel to Israel when I visited, and then her friendships across three generations at Gen X, with Ricks and her fellow volunteers, with Brissett, and with the children. Life is pretty good for her right now.

For most of us, life actually improves after the age of fifty. "The most surprising thing is that age tends to work in favor of happiness, other things being equal," Jonathan Rauch, author of *The Happiness Curve* told the *Guardian*.[7] That conclusion is based on a now-famous 2008 study of psychological well-being over the entire life course. The study of 500,000 Americans and Western Europeans, which also pulled data from other parts of the world, uncovered a *U*-shaped curve of happiness—hence the title of Rauch's book. The nadir for "the typical individual" comes in middle age—right around forty-six—and then there's an upswing. The researchers found only a little evidence for a later flattening, and a turn down, toward the end of a person's life.[8] (This is why the group of Brooklyn women, all of us now over fifty, with whom I began a tradition of a regular dinner on my first visit back from Hong Kong, is called the Upswing Club.)

The study authors tentatively propose a few possible explanations for this *U* shape. One is that by the middle of life we come to accept and adapt to our strengths and weaknesses and "quell [our] infeasible aspirations." (Better get on that.) Another possibility is that "cheerful people live longer than the miserable," which would mean that a selection effect is at work. And finally, we may, late in life, come to see that, frankly, things could be worse. Friends begin to pass away and we realize we should count our blessings for the years that remain.

Counting blessings makes sense to psychologist Laura Carstensen, founding director of the Stanford Center on Longevity. She has spent her career preaching a revised vision of the later years of life. In the early 1990s, Carstensen published an influential theory focused on time and how we want to spend it. When time is of the essence, she argues, the motivation to derive emotional meaning from life increases. What that means for social relationships, she found, is that while social networks do get smaller as we age, that narrowing is in large part intentional. People choose to spend time with those they really care about: they choose quality over quantity. It is people from the outer circles who are pared away. Furthermore, while family members are most likely to fill that inner circle, nonkin—friends—regularly do so in the absence of family. Again, what matters is the tenor of the relationship, not its origin. With fewer professional and family obligations, there are more hours for the things we want to do and the people with whom we want to do them.[9]

As you may have guessed, this pattern is repeated in many nonhuman primates. Julia Fischer is another primatologist who worked early on with Robert Seyfarth and Dorothy Cheney in Moremi. She now heads the Cognitive Ethology Lab at the German Primate Center in Göttingen and she studied aging in Barbary macaques. Over lunch at a conference on baboons, she tells me about it.

Aging monkeys have an increasing preference for predictability, she says. "They have strong bonds when they're young and then they become more selective. They have fewer partners but with these fewer partners they maintain strong bonds into old age."

It is obvious when you're watching just how powerful those bonds can be.

"It's so sweet when you see them hugging each other. It resonates, this ancient connectedness," Fischer says.

"What can humans take away from this work?" I ask.

She throws back her head and lets out a rich laugh.

"Hug each other more."

I laugh, too. Could it really be so simple?

"The vast majority of work done on social behavior shows you how important bonds are and how important body contact is and actually sitting together on the sofa. We need to preserve the time to do that. To sit together and read a book to your child or grandchild," Fischer says. "It's old primate heritage, the whole bonding issue."

Still, you have to have someone handy to hug. The compounded risk of loneliness in old age cannot be ignored. Once people retire, they lose regular interaction with colleagues. Most diseases, and the probability of getting them, worsen with age. Mental and physical capacities may diminish, and social lives may be limited by hearing loss or reduced mobility. With increasing age, it's more likely you will lose a spouse and friends start to die as well. Or, like my mother, they disappear into dementia. Heartbreaking as my mother's situation is for me, it is hugely painful for her friends, too. A person with severe dementia is still alive, but the relationship you had with them is not. My mother's closest friend, who has known her since they were newlyweds living across the street from each other in the early 1960s, lamented to her own daughters: "I thought we were going to grow old together."

Even if older people are not all automatically likely to be lonely, the indisputable fact is that there will be more of them and that the absolute number who are lonely will be larger. Demographics are just not working in favor of the fight against loneliness. For the first time in U.S. history, older adults are projected to outnumber children by 2035. Because of drops in marriage and childbearing, more of those older adults will be unmarried and childless than ever before. The percentage of older adults living alone rose steadily through the twentieth century although, according to the Pew Research Center, it decreased slightly beginning in 2014 and now hovers at 26 percent.[10]

In 2017 a British nonprofit called the Campaign to End Loneliness launched a short film on YouTube that made it clear just how serious the problem of loneliness can be.[11] The video begins with a young man alone in his apartment. The premise is this: *One week. No friends. No phone. No contact. Can you go it alone?* The young man, bearded and blue-eyed and probably in his late twenties or early thirties, has only himself for company—he can watch television, read, and otherwise entertain himself.

He records video diaries along the way and starts out cheery. "It's been a bit lonely but not unbearable." Days pass. The man naps on the couch, watches television, cooks himself a meal. In one video diary, after hearing the neighbors talking and getting ready to go out, he looks like he is going to cry. "I'm starting to find it quite difficult to be honest." He has trouble sleeping. "It's like I'm not mentally prepared for bed because there's not been a beginning, a middle and an end. It's just constant nothingness."

At the end of the week, the young man, now revealed as Joe, is finally sprung from his apartment and the real point of the exercise arrives. Joe goes next door to visit an elderly widower named Barry. A caption flashes on the screen: "More than half a million older people go up to a week without seeing anyone."

At this point in the video, I was distracted by the comment thread. It was worrisome on several fronts. For one thing, the comments were full of people who said, "This is my life." But there were also plenty of commenters who thought Joe just needed to pull himself together and didn't know how to entertain himself, or that the problem was he was addicted to technology. The introverts of the internet rejected the labeling of this week of solitary time as "loneliness."

But loneliness, as we have seen repeatedly, is the mismatch between what people want and what they get, socially. It's clear that Barry, the widower, wants more. He has been laid low by the loneliness that set in after he lost his wife two and a half years earlier. "It's not something you think about until it happens to you," he tells Joe. "Not having someone to talk to, to hug, to share things with. . . . It can be devastating." Barry's request, his call to action, is that people be aware that there are lonely folks out there. "It requires a little bit of effort, doesn't it," he says. (If you need a reminder to call your mother or grandfather, or to check on your neighbor, let this be it.)

It is worth reiterating also that not only the old get lonely. Statistics vary but combing through a variety of recent surveys it is fair to say that, at any given time, 20 percent of the population feels lonely enough that it's a problem in their lives.[12] A few recent studies have revealed, much like Teresa Seeman's study did, that specifics of our social needs vary at different points in our lives. A 2016 study, conducted by researchers at the University of North Carolina, reanalyzed data from four large, long-standing studies capturing different life stages. In both adolescence and old age, having friends was associated with a lower risk of physiological problems, and the more friends you had, the lower the risk. In other words, there was a dose–response relationship. By contrast, adults in middle age were less affected by variation in how socially connected they were. But the quality of their social relationships—whether friendships provided support or added strain—mattered more.[13] And in 2017, William Chopik of Michigan State University surveyed more than 270,000 adults and found that valuing family was important but valuing friendships became increasingly important with age. Those who did so had higher levels of health, happiness, and subjective well-being across the life span.[14]

Putting a finger on what works to build connection is not easy. Now that we know how important social integration is, many people are trying to do something to foster it, especially for lonely older adults. Ambitious efforts have started up everywhere, from local neighborhood drop-in centers to national awareness campaigns. In the United Kingdom, there is even now a Minister of Loneliness! But

unfortunately, not every program has been as successful as Generation Xchange. "We don't have awesome solutions for loneliness," says
Steve Cole. The results of intervention programs are as mixed as social
media studies. Psychologist Julianne Holt-Lunstad of Brigham Young
University, whose 2010 meta-analysis of more than 300,000 people
revealed the critical fact that social connection decreased risk of mortality by 50 percent, has spent several years working on a similar analysis of the intervention programs that have been established so far. To
her frustration, there is no clear story emerging about what works and
what doesn't.[15]

There are several potential explanations for why interventions are
tricky to pull off. One is that the importance of social relationships
has been established based on people's existing close relationships, yet
many of these interventions are with complete strangers. That works at
Generation Xchange but might not where a larger purpose is absent.
Another possibility is that interventions increase social contact, which
can help, but don't address the root of the problem: that you can still
be lonely when there are other people around if you don't feel fully connected to them. That was one of the limitations that John Cacioppo saw
with helplines set up so the elderly have someone to call. There have
been poignant stories in the press about such efforts, which clearly do
serve some purpose. One elderly British woman called every hour to
ask the time because otherwise she had no one to speak to for days on
end.[16] That's better than nothing, but unlikely to reach down into that
lonely woman's cells to boost her immune system.

More significantly, says Holt-Lunstad, most interventions are coming too late in life. "From a public health perspective, there is primary
prevention, secondary and tertiary," she says. "Most of the [loneliness]
efforts are at this tertiary level where it's already advanced." That is a
strategy that rarely works, for any public health issue. Instead, greater
emphasis needs to be put on prevention and identifying people at risk
early on. "Given that we don't necessarily know what works at this
point, one of the things I've been thinking about is really making this

related to other lifestyle-related risk factors," Holt-Lunstad says. "Habits we form early on have chronic effects over time and we often don't see the effects of them until we're older." She is calling for consensus guidelines, just as there are for nutrition, exercise, and sleep. In her imagination, there could be an app that buzzes you when you haven't called your mom in a while. Or in the same way that we make time in our schedules to exercise, we might need to make time in our schedules to see friends.

Epidemiologist Lisa Berkman agrees that in many cases, attention is being paid to relationships far too late. She reiterates just how hard it is to change behaviors that have existed for decades. "It's not very different from smoking. If you start smoking when you're 14 and stop smoking when you're 65, in many ways, the damage is done," she points out. "It's not undoable. Stopping does make some things better. It's worth doing but it's very late in the game." Very few people ever consider social relationships in the same frame as smoking. They should, says Berkman. "What you want to be thinking about is what to do in early life, or mid-life or during your working years that would really help maintain the kinds of relationships that we think are so important. That's where our work has been leading us."[17]

In search of clues for what promotes longevity, scientists and the media have explored the so-called Blue Zones—exceptional spots around the world where more people live to be one hundred than anywhere else. Social integration and connection feature prominently in these areas. When psychologist and author Susan Pinker visited one Blue Zone, a Sardinian mountain village, she found that social isolation was an impossibility. "Every centenarian we met was surrounded by a tight web of kith and kin," she writes in *The Village Effect* and then goes on to describe one woman in particular. "Zia Teresa at one hundred was what she had always been: a gregarious woman with many friends, relations and neighbors, all of whom popped in regularly to visit and chat, all the while bolstering her importance their lives, and her place in the community."[18]

As Pinker notes, you don't need to live in a literal village like Teresa, but it helps to create the feeling of one by surrounding oneself with like-minded people. Early research by Berkman and Seeman said as much. They found that geographic proximity to close friends and family is what really matters. And Nicholas Christakis and colleagues from his early days at Harvard found that widows and widowers who choose to live in neighborhoods full of other widows and widowers live longer.[19]

The hope is that the new, deeper understanding of friendship as a biological and evolutionary phenomenon will help put the importance of social connection front and center in public health circles. "When a phenomenon that we are very concerned about in humans has evolutionary roots, it has huge consequences for how we think we're going to fix the problems that arise from it," says primatologist Susan Alberts, whose work in the baboons of Amboseli has contributed dramatically to the conversation. If friendship were purely cultural, a construct of human society, promoting it and avoiding its opposite—loneliness—would be a cultural matter. But if baboons, macaques, dolphins, and even prairie voles and zebra fish have—and need—friends, then the answers lie elsewhere. If something has the power to change the way our immune systems work, to increase or decrease the surges of neurotransmitters in our brains, to extend or cut short our lives, then it must be dealt with as such.

Holt-Lunstad has not been shy about declaring how critical it is that social connection and loneliness be regarded as a public health issue. "For too long, this has been regarded as essential to emotional well-being but not necessarily physical well-being and longevity," she told me when we met in her office at Brigham Young with mountains filling the window behind her. "People view relationships as something that is personal and emotional, not so related to the body." She has made a point of trying to get her work published in top medical journals, not just psychology journals, because she wants health professionals to hear her message. In a call-to-arms paper in 2017, she declared that American health organizations have been slow to recognize the substantial evidence that has piled up showing that social connection reduces mortality, while a lack of it significantly increases risk, even more than the other issues that do rate attention and resources, such as obesity and air pollution.[20]

Communities, institutions, and corporations will have to act if there is to be real change. Telling individuals how to alter their behaviors and habits doesn't usually work. Doctors and public health experts have been advising people for years to exercise and eat vegetables. Not everyone has followed that advice. But what if friendship and social connection were a consideration for policy makers? "If our goal is not to invent new relationships necessarily, but to help foster the kinds of natural relationships people have in their lives and not be destructive of them, what would we be doing?" Lisa Berkman asks. "We would have family friendly policies. We would not move people across the country so that there's a lot of geographic dislocation. We would encourage voluntary activities, and community cohesion, and neighborhoods [where] people are naturally likely to be in contact with each other. It's those kinds of policies, at either a company level, or a state, or city level that actually have the biggest impact on social isolation."

Anyone who grew up in a military family, like my friend Stephanie, can appreciate the importance of policy makers bringing some awareness of the need for social bonds to their thinking. Stephanie attended at least ten schools before college and moved more times than that, sometimes in the middle of a school year. Her childhood was one long stretch of being the new kid with no lasting friendships.

Another kind of example comes from the health care system, which must acknowledge the possibility that friends can provide meaningful support to patients. Former English professor Joan DelFattore, who is in her seventies, has written and spoken extensively about her own experience of being "sick while single." An oncologist, discussing what treatment to give her, asked whether she had a spouse or children. "When I said no to both, he looked genuinely concerned," she wrote in the *Washington Post*.[21] "'But how will you manage?' he asked." In the absence of immediate family to help DelFattore through the treatment, the doctor proposed to give her only one mild drug, although the standard of care was a much harsher—and more effective—combination chemotherapy. "When I tried to describe my strong network of friends and extended family, he talked right over me," she wrote. DelFattore changed doctors and got the combination chemotherapy, which she credits with saving her life. As an experienced researcher, albeit not in the medical world, she carried out a review of every study she could find about the differences in treatment for single people. "Study after study reported significant differences in treatment rates between married and unmarried patients," she writes. DelFattore hopes that sharing her story will help shift medical focus to the support a patient needs, rather than the specific roles of the people who might provide it.

It's also clear that social isolation is a financial issue. The lack of social contacts among older adults costs Medicare $6.7 billion a year, mostly from spending on nursing facilities and hospitalization for those with no network. "The effect of isolation is extraordinarily powerful," Donald Berwick, former administrator of the Centers for Medicare and Medicaid Services, told the *Wall Street Journal*. "If we want to achieve health for our population, especially vulnerable people, we have to address loneliness."[22]

Hard as it is for individuals to change our habits, we are not absolved of responsibility. We must make friendship a priority and factor it in to the way we plan our time—and our children's time. Yes, you can choose your friends, but you must also more generally choose friendship—embrace it, invest in it, work at it. Put time and attention into building quality relationships. Be mindful of your social convoy. You cannot afford not to.

Steve Cole even imagines a future in which we might use biological feedback to get a read on our personal well-being. "If this gene expression profile measures something about our body's opinion of whether we're doing well in life, we might be able to take that as a metric and use it to do little experiments on ourselves—changing our lifestyle in various ways to see which ones improve our molecular well-being, and hopefully do that long before it precipitates a disease." That would be just the thing to tell the difference between those who are truly "satisfied" with their level of social connection and those who yearn for more, those who are deeply connected and those who are not. Although he has not yet been able to explicitly test the gene expression of those who feel lucky in friendship, Cole believes there is logic in this possibility. "Humans are incredibly, fundamentally social. You almost never find a lone human being in the world. This brain we have for communicating, networking and coordinating, that's really the secret sauce in the human life history strategy. It makes sense that that stuff also helps us feel well and safe and secure and it is exactly the psychology that knocks down threat and uncertainty."

Evidence that this is so comes from a handful of studies that have successfully followed humans for their entire lives. That is harder to do in people than in baboons and macaques because of our much longer life spans, as well as our tendency to move or just stop responding to the entreaties of scientists. But a few persistent efforts have paid off handsomely.

The most exhaustive and long-running of these studies began at Harvard University in 1937.[23] Arlie Bock, "a brusque, no-nonsense physician" who ran health services at Harvard, dreamed up the project. He began with 268 carefully selected sophomores and conducted thorough medical exams, home visits to parents with social workers to collect life histories, and extensive psychological evaluations. In 1938 a second group of men was added who were far from the Harvard men in every sense. These teenagers (they were eleven to sixteen) came from the poorest neighborhoods in Boston. Every few years, the researchers followed up with more exams and more questions. In total, 724 men were followed for their lifetimes over more than seventy-five years.

Bock's original grand intention was to focus not on ill health but on what it took to live well. And decades later, the answer was quite clear. In 2008 longtime study director George Vaillant was asked, "What have you learned?" His response was emphatic: "That the only thing that really matters in life are your relationships to other people."

The current study director is Robert Waldinger. Late in November 2015, Waldinger echoed Vaillant in a TEDx Talk that has since accumulated nearly thirty million views. "The clearest message we get from this 75-year study is this: Good relationships keep us happier and healthier. Period."[24]

Waldinger backed up his statement with three main lessons. First, that social connections are really good for us and that loneliness kills. If you've gotten this far in this book, you know exactly why this is true and how we know it. The second lesson is that quality matters as much or more than quantity, which we have also seen. "Conflict is bad for your health, good warm relationships are protective," Waldinger said. And here's where the Harvard study really adds to the story. They took the men who lived into their eighties and looked back to the data from the middle of their lives to see whether it would predict "who was going to live to be a happy, healthy octogenarian and who was not." Waldinger and his team gathered everything they knew about the men at age fifty. "It wasn't their middle-aged cholesterol levels that predicted how they were going to grow old. It was how satisfied they were in their relationships. The people most satisfied at 50 were the healthiest at age 80. Good close relationships seem to buffer them from the slings and arrows of getting old." Those relationships even made physical pain more bearable.

The third lesson is that good relationships don't just protect our bodies, they protect our brains. Those who were securely attached to another person in their eighties stayed cognitively sharper for longer. "Those relationships don't have to be smooth all the time," Waldinger stressed. "They could bicker day in and day out, but as long as they felt they could count on each other when the going got tough, those arguments didn't take a toll on their memories."

Strikingly, Waldinger noted, while some of the details are new, the wisdom that relationships are good for us is not new. "It's as old as the hills," he said. So, "why is this so hard to get and so easy to ignore?" Why, in other words, can the critical importance of friendship and strong social bonds have been hiding in plain sight for so very long?

In Waldinger's view it's because we humans like a quick fix. "Relationships are messy and complicated, the hard work of tending to family or friends is not sexy or glamorous. It's also long. It never ends." The people in his study who were happiest in retirement had worked at it. They valued and tended to their relationships. They actively worked to replace coworkers with new playmates. They put in the time.

The lesson for all of us can be even simpler than Waldinger's three points. It's to take all this carefully gathered information on how deeply wired we are for connection and then, as Steve Cole likes to say: "Plan your day accordingly."

꒰

Ever since I started working on this book, my boys have reveled in the ability to remind me about the importance of friends. When I started, Jake was graduating from high school. As I finished, Matthew was a senior. Just before the start of the school year, he wanted to spend the last weekend of the summer away with friends. They were tightly bonded, these boys. So much so that one of them later gave his senior speech—his five-minute opportunity to tell the school community what mattered to him—about his friends, beginning with Matty.

On this particular weekend, though, I thought Matthew needed to work on his college applications and the summer homework that wasn't yet finished. In his usual fashion, he had put those things off to the last minute, lured by the joys of summer. I thought there was a lesson in there about the consequences of that decision. But then he reminded me that this was the last year he would have to spend with these friends in this way. Who knew where everyone would be the following September? He was already storing up the experience just as his brother Jake and his friend Christian had done.

Reader, I let him go.

Acknowledgments

Where else to begin but with my friends? The sad irony is that the effort of reporting and writing this book about friendship did, in fact, take me away from my friends, though I tried to not let that happen. Even so, the members of my inner and outer circles were there for me through this book's long gestation—supporting me, asking useful questions, weighing in with their own stories. Moira Bailey and Stephanie Holmes served as early readers. Leah MacFarlane and Suzanne Myers urged me on, week in and week out, as I wondered whether I would ever solve the various problems that presented themselves or manage to write the thousands of words that still lay ahead, and then cut thousands of words when it turned out I had overdone it. Many others provided levity and/or succor as and when I needed either. Thank you especially to Amy and Tom Jakobson and Elizabeth Schwarz, Jenny McMahon, Dan Dolgin, and the members of the Upswing Club (Stephanie and Moira, Susanna Wenniger, Shirley Hedden, Lexy Lovell, Sarah Prud'homme, and Claire Wright) for so many sustaining conversations and so much support. Thank you to the rest of my Brooklyn and Berkeley Carroll friends for demonstrating daily the importance of community. And thank you to my high school and college friends for reminding me of the value of persistent relationships.

I was also fortunate to make new friends through my years of work on this book, chief among them Lauren Brent, Robert Seyfarth and Dorothy Cheney, Michael Platt, and Susan Alberts. Their wisdom and willingness to talk to me again and again was essential, and also fun. When Dorothy died in November 2018, I felt the loss of the possibility of a deeper friendship but was so grateful for the short time we had together. And I must extend a special note of thanks to Robert, who served as a scientific advisor as well as a friend, read every word multiple times and made the book immeasurably better every step of the way. Any remaining errors or misjudgments are mine alone.

I extend heartfelt thanks to the many sources who gave me their time. Many appear in this book but there are quite a few who do not. There is far more good work out there on friendship than what I was able to include here due to the limits of space and time. But every conversation I engaged in, talk I listened to, and book and paper I read helped to inform and deepen my understanding. It is difficult to thank everyone I have spoken to, but I will try.

On Cayo Santiago, thanks to Michael Platt and Lauren Brent, but also Giselle Caraballo, Angelina Ruiz Lambides, Laurie Santos, Lindsay Drayton, Sam Larson, Aparna Chandrashekar, Michael Montague, Dario Maestripieri, Amaury Michel, Richard Rawlins, Matthew Kessler, and everyone at the Caribbean Primate Research Center who made my visits possible. Cayo Santiago was directly in the path of Hurricane Maria and the island was devastated by the storm, as was the surrounding area. Yet the monkeys somehow survived. I salute the resilience and determination of the staff of the CPRC. Their own lives were dramatically altered by Maria. They have worked in tremendously difficult conditions, with the help of dedicated scientists and donations from around the world, to care for the animals and begin to restore the island to what it was.

At Amboseli, thanks to Susan Alberts, Jeanne Altmann, Beth

Archie, Jenny Tung, Raphael Matutua, Kinyua Waretere, and everyone else at the Project and at the Tortellis Lodge who made our visit
so special. Thanks to Joan Silk, Robin Dunbar, Julia Ostner, Oliver
Schulke, Julia Fischer, and Federica Dal Pasco for sharing wisdom
about baboons and monkeys. Thanks to John Capitanio, Brenda
McCowan, and Tamara Weinstein at UC Davis and Steve Cole at
UCLA. Thanks to Jim House and Toni Antonucci for welcoming me
to the University of Michigan, Nicholas Christakis for welcoming me
to Yale, and Claude Fischer for welcoming me to his home. Thanks to
Lisa Berkman, Mark Leary, Mario Small, Frans de Waal, Jamil Zaki,
Carolyn Zahn-Waxler, Eric Nelson, Jaana Juvonen, Larry Steinberg,
Jeff Hall, Beverley Fehr, and Bill Chopik for their time. And thanks to
Dan Hruschka for providing inspiration and so very many resources in
his book. Thanks to Francis McGlone and Håkan Olausson for intro-
ducing me to affective touch. Thanks to Sarah Lloyd-Fox, Laura Piz-
zarolli, and Mark Johnson at Birkbeck, and Thalia Wheatley, Carolyn
Parkinson, Adam Kleinbaum, Emma Templeton, and everyone in the
Wheatley lab, and Joy Hirsch for her discussion of brain coherence.
Thanks also to Julianne Holt-Lunstad, Bert Uchino, Gary Bernt-
son, Louise Hawkley, Wendi Gardner, Bill Patrick, and Jenny de Jong
Gierveld. Thanks to Megan Gunnar, Dylan Gee, and Dan Abrams. At
the Platt lab, thanks to Michael Platt, Geoff Adams, Yao Jiang, Heidi
Steffen, Naz Belkaya, and everyone who welcomed me and explained
just what it was they were doing. Thanks to Nancy Kanwisher and
Margaret Livingstone for letting me corral them by the posters. And
thanks to Keith Hampton, Ariel Shensa, Andrew Przybylski, Candice
Odgers, Jeff Hancock, Melissa Hunt, and Tracy Dennis-Tiwary for
their wisdom on social media.

    And finally, my heartfelt thanks to the wonderful people of the 74th
Street School Generation Xchange program for sharing their stories.
Teresa Seeman, D'Ann Morris, Paula Dutton, Linda Ricks, Theresa
Brissett, Michael Travers, Bertha Wellington, Menora Garner, Ber-
nice Livingston, Barbara Phillips, Wynona Prince, Barbara Bass, Pat

Templeton, and Leatrice Jones, you gave me the perfect ending for this book.

Many of those named above read chapters and continued to provide wisdom. Again, any remaining mistakes or judgments are mine alone.

I am grateful to my past and present editors at *Scientific American* and *Scientific American Mind*—especially Seth Fletcher, Gary Stix, Claudia Wallis, Kristin Ozelli, Ingrid Wickelgren, and Daisy Yuhas—for giving me space to write about friendship, touch, empathy, baboons, and so much more as I explored the ideas in this book.

Lucky for me, my agent Dorian Karchmar is also a friend. Her belief in me and the project kept me going. I am thankful to my editor Quynh Do at W. W. Norton who was so enthusiastic from the start. And heartfelt thanks to Doron Weber and the Alfred P. Sloan Foundation for their generous support.

Finally, thank you to my family. To Nancy Redmond and Lance Herning for friendship and support through some difficult years. To my sons Jacob, Matthew, and Alex for once again allowing me to write about them, and to Christian Denis, my fourth son, for the same. And to my husband, Mark Justh, who has been my best friend even though we have had our share of ambivalent moments. After more than thirty years together, our bond is long lasting, positive, and equitable. I look forward to being eighty together. You are all part of my inner circle, the people without whom I cannot imagine life.

Notes

Introduction: A New Science

1. Details of the history of Cayo Santiago are taken from multiple interviews and several historical sources. Most important were the articles written by Matt Kessler and Richard Rawlins. See Matthew J. Kessler and Richard G. Rawlins, "A 75-Year Pictorial History of the Cayo Santiago Rhesus Monkey Colony," *American Journal of Primatology* 78, no. 1 (2016): 6–43; Richard G. Rawlins and Matt J. Kessler, "The History of the Cayo Santiago Colony," in Richard G. Rawlins and Matt J. Kessler, eds., *The Cayo Santiago Macaques: History, Behavior, and Biology* (SUNY Press, 1986). I also relied on Donna J. Haraway, "A Semiotics of the Naturalistic Field: From C. R. Carpenter to S. A. Altmann, 1930–1955," in *Primate Visions: Gender, Race, and Nature in the World of Modern Science* (Routledge, 2013).

2. Jackie Buhl, in *Primates de Caraïbe* (2013), a documentary by Jean Christophe Ribot and Jack Silberman.

3. I conducted multiple interviews with Lauren Brent. Several were on Cayo Santiago in May 2016. We also spoke on the phone, at conferences, at Yale University in November 2015, and at the University of Exeter in December 2017.

4. See the BRAIN Initiative, https://www.braininitiative.nih.gov.

5. "short shrift": Beverley Fehr, *Friendship Processes*, vol. 12 (Sage, 1996), chap. 1; Joan B. Silk, "Using the 'F'-Word in Primatology," *Behaviour* 139, no. 2–3 (2002): 421. Also interviews with Fehr, Silk, Robert Seyfarth, Nicholas Christakis, and many others.

6. Clive Staples Lewis, *The Four Loves* (Houghton Mifflin Harcourt, 1991).

7. Robert M. Seyfarth and Dorothy L. Cheney, "The Evolutionary Origins of Friendship," *Annual Review of Psychology* 63 (2012): 153–77.

8. Ana I. Faustino, André Tacão-Monteiro, and Rui F. Oliveira, "Mechanisms of Social Buffering of Fear in Zebrafish," *Scientific Reports* 7 (2017): 44329.

9. Keith M. Kendrick et al., "Sheep Don't Forget a Face," *Nature* 414, no. 6860 (2001): 165.

10. Frans De Waal, *Are We Smart Enough to Know How Smart Animals Are?* (W. W. Norton & Company, 2016), 25.

11. Presence of a friend . . . : Roman M. Wittig et al., "Social Support Reduces Stress Hormone Levels in Wild Chimpanzees across Stressful Events and Everyday Affiliations," *Nature Communications* 7 (2016): 13361; Simone Schnall et al., "Social Support and the Perception of Geographical Slant," *Journal of Experimental Social Psychology* 44, no. 5 (2008): 1246–55; James A. Coan, Hillary S. Schaefer, and Richard J. Davidson, "Lending a Hand: Social Regulation of the Neural Response to Threat," *Psychological Science* 17, no. 12 (2006): 1032–39.

12. See chapter 9.

13. Sandra Manninen et al., "Social Laughter Triggers Endogenous Opioid Release in Humans," *Journal of Neuroscience* 37, no. 25 (2017): 6125–31.

14. I visited Cayo twice in July 2015 and May 2016, and interviewed eight researchers, assistants, and caretakers.

15. In addition to the various histories of Cayo that detail how it is set up, the arrangements are described in the many scientific papers that have arisen from the work such as K. K. Watson et al., "Genetic Influences on Social Attention in Free-Ranging Rhesus Macaques," *Animal Behaviour* 103 (2015): 267–75.

16. Eva Hagberg Fisher, *How to Be Loved: A Memoir of Lifesaving Friendship* (Houghton Mifflin Harcourt, 2019). Reviewed in the *New York Times Book Review*, March 3, 2019.

17. Theodore M. Newcomb, *The Acquaintance Process* (New York: Holt, Rinehart & Winston. 1961).

Chapter One: Fierce Attachment

1. The history of the relationship between Bowlby and Hinde is drawn from several sources: Frank C. P. Van Der Horst, Rene Van der Veer, and Marinus H. Van Ijzendoorn, "John Bowlby and Ethology: An Annotated Interview with Robert Hinde," *Attachment and Human Development* 9, no. 4 (2007): 321–35; Patrick Bateson, Joan Stevenson-Hinde, and Tim Clutton-Brock, "Robert Aubrey Hinde CBE, 26 October 1923–23 December 2016," *Biographical Memoirs of Fellows of the Royal Society* 65 (2018): 151–77; Jane Goodall et al., "Remembering My Mentor: Robert Hinde," www.janegoodall.org.

2. Van Der Horst et al., "John Bowlby," 325.

3. John Bowlby, *Forty-Four Juvenile Thieves: Their Characters and Home-Life* (London: Ballière, Tindall & Cox, 1946).

4. Michael Pakaluk, ed., *Other Selves: Philosophers on Friendship* (Hackett, 1991).

5. See also Alexander Nehamas, *On Friendship* (Basic Books, 2016).

6. Nehamas, *On Friendship*, 62.

7. Pakaluk, *Other Selves*, 1.

8. Pakaluk, *Other Selves*, 34. For a discussion of Aristotle's philosophy of friendship see Nehamas, *On Friendship*, chap. 1.

9. Nehamas, *On Friendship*, 51.

10. William James, *Letters of William James* (Atlantic Monthly Press, 1920), 109.

11. Émile Durkheim, *Suicide: A Study in Sociology* (Free Press, 1897/1997).

12. John Watson: Deborah Blum, *Love at Goon Park: Harry Harlow and the Science of Affection* (Basic Books, 2002/2011), 37–40.

13. René A. Spitz, "Hospitalism: An Inquiry into the Genesis of Psychiatric Conditions in Early Childhood," *The Psychoanalytic Study of the Child* 1, no. 1 (1945): 53–74.

14. René A. Spitz, "Hospitalism: A Follow-Up Report on Investigation Described in Volume I, 1945," *The Psychoanalytic Study of the Child* 2, no. 1 (1946): 113–17.

15. Blum, *Love at Goon Park*, 52.

16. Blum, *Love at Goon Park*, 53.

17. The influence of ethology and Robert Hinde on Bowlby: "John Bowlby and Ethology: An Annotated Interview with Robert Hinde," *Attachment and Human Development* 9, no. 4 (2007); Sarah Blaffer Hrdy, *Mother Nature* (Random House, 1999), chap. 16; John Bowlby and World Health Organization, "Maternal Care and Mental Health: A Report Prepared on Behalf of the World Health Organization as a Contribution to the United Nations Programme for the Welfare of Homeless Children." *Bulletin of the World Health Organization* 3 (1951): 355–534.

18. Melvin Konner, *The Evolution of Childhood: Relationships, Emotion, Mind* (Belknap Press, 2010), 70–72; John Alcock, *The Triumph of Sociobiology* (Oxford University Press, 2001), 94; Richard W. Burkhardt, *Patterns of Behavior: Konrad Lorenz, Niko Tinbergen, and the Founding of Ethology* (University of Chicago Press, 2005); the lives and achievements of all three are described in Dale Peterson, *Jane Goodall: The Woman Who Redefined Man* (Houghton Mifflin Harcourt, 2014), chap. 19.

19. Karl Von Frisch, *The Dance Language and Orientation of Bees* (Harvard University Press, 1967).

20. Nikolaas Tinbergen, *The Study of Instinct* (Clarendon Press, 1951); Patrick Bateson and Kevin N. Laland, "Tinbergen's Four Questions: An Appreciation and an Update," *Trends in Ecology and Evolution* 28, no. 12 (2013): 712–18.

21. Konrad Lorenz, "Companions as Factors in the Bird's Environment," *Studies in Animal and Human Behavior* 1 (1970): 101–258; Hrdy, *Mother Nature*, 197 (photo).

22. Hinde studies summarized in John Bowlby, *Separation: Anxiety and Anger* (Basic Books, 1973), 60–74; see, for example, Robert A. Hinde and Yvette Spencer-Booth,

"Effects of Brief Separation from Mother on Rhesus Monkeys," *Science* 173, no. 3992 (1971): 111–18.

23. Hinde as a mentor is described in Peterson, *Jane Goodall*, chap. 19, and Goodall et al., "Remembering My Mentor."

24. Van Der Horst et al., "John Bowlby."

25. Blum, *Love at Goon Park*, both quotes taken from the Introduction to the 2011 edition.

26. John Bowlby, "The Nature of the Child's Tie to His Mother," *International Journal of Psychoanalysis* 39 (1958): 350–73.

27. John Bowlby, *Attachment* (Basic Books, 1969/1982), 180.

28. Bowlby, *Attachment*, 207–8.

29. Sources for the history of Altmann's work on Cayo are Edward O. Wilson, *Naturalist* (Island Press, 2013), chap. 16; *Lord of the Ants* (Nova/WGBH by Windfall Films and Neil Patterson Productions, 2008); Haraway, *Primate Visions*, chap. 5.

30. "We're 95% sure . . .": Altmann speaking in *Lord of the Ants*.

31. "I had to have one kind of animals . . .": Wilson, *Naturalist*, chap. 1, loc. 187, Kindle. This memoir also provided details of Wilson's boyhood.

32. Wilson, *Naturalist*, chap. 10, loc. 1955.

33. Carl Zimmer, *Evolution: The Triumph of an Idea* (Harper Collins, 2001), 86. The book also served as a major source for this brief history of thinking on natural selection and heredity.

34. Zimmer, *Evolution*, 93.

35. In addition to Zimmer's book, I relied on Wilson, *Naturalist*, chap. 7, for this history.

36. Altmann in *Lord of the Ants*.

37. Details of the history of Cayo Santiago are taken from multiple interviews and several historical sources. Most important were the history articles written by Kessler and Rawlins, "A 75-Year Pictorial History"; Rawlins and Kessler, "History of the Cayo"; and Haraway, *Primate Visions*.

38. Carpenter's account of the establishment of Cayo is reprinted in Rawlins and Kessler, "History of the Cayo," 14–21.

39. Quoted in Haraway, *Primate Visions*, 93.

40. Stuart A. Altmann, "A Field Study of the Sociobiology of Rhesus Monkeys, *Macaca mulatta*" (Annals of the New York Academy of Science, 1962).

41. Wilson, *Naturalist*, chap. 16, loc. 3884.

42. Wilson, *Naturalist*, chap. 16, loc. 3897.

43. *Lord of the Ants*.

44. "What, we are compelled to ask . . .": E. O. Wilson, *Sociobiology: The New Synthesis* (Harvard University Press, 1975/2000), 1.

45. Wilson, *Naturalist*, chap. 16, loc. 3964.

46. William D. Hamilton, "The Genetical Evolution of Social Behaviour I," *Journal of Theoretical Biology* 7, no. 1 (1964): 1–16.

47. "terrible public speaker . . .": Recounted in Robert Trivers, *Natural Selection and Social Theory: Collected Papers of Robert Trivers* (Oxford University Press, 2002), 10.

48. Wilson recounts his conversations with Hamilton in *Naturalist*, chap. 16.

49. George C. Williams, *Adaptation and Natural Selection* (Princeton University Press, 1966).

50. Robert L. Trivers, "The Evolution of Reciprocal Altruism," *Quarterly Review of Biology* 46, no. 1 (1971): 35–57.

51. Wilson, *Sociobiology*, chap. 1, loc. 542, Kindle.

52. "By comparing . . .": Wilson, *Sociobiology*, chap. 27, loc. 17102.

53. Elizabeth Allen et al., "Against 'Sociobiology,'" *New York Review of Books*, November 13, 1975.

54. Boyce Rensberger, "Sociobiology: Updating Darwin on Behavior," *New York Times*, May 28, 1975, 1.

55. "Why You Do What You Do," *Time*, August 1, 1977.

Chapter Two: Building a Social Brain

1. Sarah Blaffer Hrdy, *Mothers and Others: The Evolutionary Origins of Mutual Understanding* (Harvard University Press, 2009), 6.

2. Konner, *Evolution of Childhood*, 89.

3. Author interview with Mark Johnson.

4. Bowlby, *Attachment*, 265.

5. The Kanwisher lab posts videos of her lectures at https://nancysbraintalks.mit.edu/video/humans-are-highly-social-species.

6. Paul D. MacLean and Vojetch Adalbert Kral, *A Triune Concept of the Brain and Behaviour* (University of Toronto Press, 1973).

7. I have drawn on many interviews, lectures, and sources to report on the social brain, such as the work of Michael Platt, Thalia Wheatley, Carolyn Parkinson, and Ralph Adolphs: for example, Sébastien Tremblay, K. M. Sharika, and Michael L. Platt, "Social Decision-Making and the Brain: A Comparative Perspective," *Trends in Cognitive Sciences* 21, no. 4 (2017): 265–76; Carolyn Parkinson and Thalia Wheatley, "The Repurposed Social Brain," *Trends in Cognitive Sciences* 19, no. 3 (2015): 133–41; Ralph Adolphs, "The Social Brain: Neural Basis of Social Knowledge," *Annual Review of Psychology* 60 (2009): 693–716.

8. Author interview with Sarah Lloyd-Fox.

9. M. H. Johnson, J. J. Bolhuis, and G. Horn, "Interaction between Acquired Preferences and Developing Predispositions during Imprinting," *Animal Behaviour* 33, no. 3 (1985): 1000–1006.

10. Mark Johnson, "Memories of Mother," *New Scientist* 117, no. 1600 (1988): 60–62.

11. Warren Jones and Ami Klin, "Attention to Eyes Is Present but in Decline in 2–6-month-old Infants Later Diagnosed with Autism," *Nature* 504, no. 7480 (2013): 427.

12. R. Jenkins, A. J. Dowsett, and A. M. Burton, "How Many Faces Do People Know?" *Proceedings of the Royal Society B* 285, no. 1888 (2018): 20181319.

13. Lisa Feldman Barrett, *How Emotions Are Made: The Secret Life of the Brain* (Houghton, Mifflin Harcourt, 2017).

14. Carolyn C. Goren, Merrill Sarty, and Paul Y. K. Wu, "Visual Following and Pattern Discrimination of Face-like Stimuli by Newborn Infants," *Pediatrics* 56, no. 4 (1975): 544–49.

15. Mark H. Johnson et al., "Newborns' Preferential Tracking of Face-like Stimuli and Its Subsequent Decline," *Cognition* 40, no. 1–2 (1991): 1–19.

16. Nancy Kanwisher TED Talk, March 2014.

17. Nancy Kanwisher, Josh McDermott, and Marvin M. Chun, "The Fusiform Face Area: A Module in Human Extrastriate Cortex Specialized for Face Perception," *Journal of Neuroscience* 17, no. 11 (1997): 4302–11.

18. Michael J. Arcaro et al., "Seeing Faces Is Necessary for Face-Domain Formation," *Nature Neuroscience* 20, no. 10 (2017): 1404.

19. "that first study": Sarah Lloyd-Fox et al., "Social Perception in Infancy: A Near Infrared Spectroscopy Study," *Child Development* 80, no. 4 (2009): 986–99.

20. Sarah Lloyd-Fox et al., "Cortical Specialisation to Social Stimuli from the First Days to the Second Year of Life: A Rural Gambian Cohort," *Developmental Cognitive Neuroscience* 25 (2017): 92–104.

21. Lise Eliot, *What's Going on in There? How the Brain and Mind Develop in the First Five Years of Life* (Bantam Books, 1999), chap. 10.

22. Sarah Lloyd-Fox et al., "The Emergence of Cerebral Specialization for the Human Voice over the First Months of Life," *Social Neuroscience* 7, no. 3 (2012): 317–30.

23. Author interview with Daniel Abrams; Daniel A. Abrams et al., "Neural Circuits Underlying Mother's Voice Perception Predict Social Communication Abilities in Children," *Proceedings of the National Academy of Sciences* 113, no. 22 (2016): 6295–6300.

24. Lydia Denworth, "The Social Power of Touch," *Scientific American Mind* 26, no. 4 (2015): 30–39.

25. Author interview with Håkan Olausson.

26. In addition to interviews with Håkan Olausson and Francis McGlone, I drew on the literature on affective touch—for example, Francis McGlone, Johan Wessberg, and Håkan Olausson, "Discriminative and Affective Touch: Sensing and Feeling," *Neuron* 82, no. 4 (2014): 737–55; Line S. Löken et al., "Coding of Pleasant Touch by Unmyelinated Afferents in Humans," *Nature Neuroscience* 12, no. 5 (2009): 547.

27. Konner, *Evolution of Childhood*, 219–20.

28. Autism: Sarah Lloyd-Fox et al., "Cortical Responses before 6 Months of Life Associate with Later Autism," *European Journal of Neuroscience* 47, no. 6 (2018): 736–49.

29. Author interview with Frans de Waal; Frans B. De Waal, "Putting the Altruism Back into Altruism: The Evolution of Empathy," *Annual Review of Psychology* 59 (2008): 279–300.

30. Lydia Denworth, "I Feel Your Pain," *Scientific American* 317, no. 6 (2017): 58–63.

31. Author interview with Jamil Zaki.

32. Frans De Waal lecture, Primate Society of Great Britain, December 2017.

33. Carolyn Zahn-Waxler et al., "Development of Concern for Others," *Developmental Psychology* 28, no. 1 (1992): 126.

34. Recounted by Rebecca Saxe at TEDGlobal 2009, "How We Read Each Other's Minds."

35. Michael Tomasello, *Why We Cooperate* (MIT Press, 2009).

Chapter Three: Friendship under the Skin

1. A 2015 study: Johannes Thrul and Emmanuel Kuntsche, "The Impact of Friends on Young Adults' Drinking over the Course of the Evening: An Event-Level Analysis," *Addiction* 110, no. 4 (2015): 619–26.

2. "Achievements in Public Health, 1900–1999: Control of Infectious Diseases," Centers for Disease Control, *MMWR Weekly* 48 no. 29 (July 30, 1999): 621–29.

3. E. Cuyler Hammond and Daniel Horn, "The Relationship between Human Smoking Habits and Death Rates: A Follow-up Study of 187,766 Men," *Journal of the American Medical Association* 155, no. 15 (1954): 1316–28. A summary of the turning point is in "The Study That Helped Spur the U.S. Stop-Smoking Movement," American Cancer Society, January 9, 2014, www.cancer.org.

4. Lisa F. Berkman and Lester Breslow, *Health and Ways of Living: The Alameda County Study* (Oxford University Press, 1983); Jeff Housman and Steve Dorman, "The Alameda County Study: A Systematic, Chronological Review," *Journal of Health Education* 36, no. 5 (2005): 302–8.

5. We Are Public Health interview with Len Syme, November 3, 2013, www.wearepublichealthproject.org.

6. I'm referring to the Framingham Heart Study, the Alameda County Study, and the Tecumseh Community Health Study.

7. John A. Napier, "Field Methods and Response Rates in the Tecumseh Community Health Study," *American Journal of Public Health and the Nations Health* 52, no. 2 (1962): 208–16.

8. See Hans Selye, *The Stress of Life* (McGraw-Hill, 1956/1976). I also drew on my interview with James House.

9. Selye, *Stress of Life*, Preface to the Revised Edition, xiii.

10. John Cassel, "The Contribution of the Social Environment to Host Resistance: The Fourth Wade Hampton Frost Lecture," *American Journal of Epidemiology* 104, no. 2 (1976): 107–23; Sidney Cobb, "Social Support as a Moderator of Life Stress," *Psychosomatic Medicine* (1976).

11. James S. House, "Work Stress and Social Support," *Addison-Wesley Series on Occupational Stress* (1983), 14.

12. Author interview with Lisa Berkman.

13. Lisa F. Berkman and S. Leonard Syme, "Social Networks, Host Resistance, and Mortality: A Nine-Year Follow-up Study of Alameda County Residents," *American Journal of Epidemiology* 109, no. 2 (1979): 186–204.

14. These comments and much of these details drawn from author interviews with James House and Debra Umberson.

15. James S. House, Karl R. Landis, and Debra Umberson, "Social Relationships and Health," *Science* 241, no. 4865 (1988): 540–45.

16. Julianne Holt-Lunstad, Timothy B. Smith, and J. Bradley Layton, "Social Relationships and Mortality Risk: A Meta-Analytic Review," *PLoS Medicine* 7, no. 7 (2010): e1000316.

17. William James, *Principles of Psychology*, vol. 2 (New York: Holt, 1890/1918), 450.

18. John T. Cacioppo and Louis G. Tassinary, "Inferring Psychological Significance from Physiological Signals," *American Psychologist* 45, no. 1 (1990): 16.

19. "shot out of a cannon": Author interview with Steve Cole. Additional background on Cacioppo, who died in March 2018, is taken from author interviews with many colleagues and from an award citation for Cacioppo, *American Psychologist* 57, no. 11 (November 2002), 817–31.

20. John T. Cacioppo and Gary G. Berntson, "Social Psychological Contributions to the Decade of the Brain: Doctrine of Multilevel Analysis," *American Psychologist* 47, no. 8 (1992): 1019.

21. Author interview with Gary Berntson.

22. Janice K. Kiecolt-Glaser et al., "Slowing of Wound Healing by Psychological Stress," *The Lancet* 346, no. 8984 (1995): 1194–96.

23. Bert N. Uchino, John T. Cacioppo, and Janice K. Kiecolt-Glaser, "The Relationship between Social Support and Physiological Processes: A Review with Emphasis on Underlying Mechanisms and Implications for Health," *Psychological Bulletin* 119, no. 3 (1996): 488.

24. Author interviews with Gary Berntson and Louise Hawkley. Also John T. Cacioppo et al., "Lonely Traits and Concomitant Physiological Processes: The MacArthur Social Neuroscience Studies," *International Journal of Psychophysiology* 35, no. 2–3 (2000): 143–54.

25. Louise C. Hawkley and John T. Cacioppo, "Loneliness Matters: A Theoretical

and Empirical Review of Consequences and Mechanisms," *Annals of Behavioral Medicine* 40, no. 2 (2010): 218–27.

26. John Cacioppo, "What's Social about Social Neuroscience?" (Keynote address, Society for Social Neuroscience, 2015).

27. Author interview with Louise Hawkley.

28. Paul Tillich, "Let Us Dare to Have Solitude," *Union Seminary Quarterly Review* (May 1957).

29. Jenny de Jong Gierveld, "A Review of Loneliness: Concept and Definitions, Determinants and Consequences," *Reviews in Clinical Gerontology* 8, no. 1 (1998): 73–80.

30. Author interview with Jenny de Jong Gierveld.

31. John T. Cacioppo and William Patrick, *Loneliness: Human Nature and the Need for Social Connection* (W. W. Norton & Company, 2008), 8.

32. Accounts of these studies come from author interviews with Gary Berntson and Louise Hawkley, from Cacioppo and Patrick, *Loneliness*, and from various journal articles.

33. This is the Chicago Health Aging and Social Relations Study, known as CHASRS.

34. Cacioppo and Patrick, *Loneliness*.

35. Laura Fratiglioni et al., "Influence of Social Network on Occurrence of Dementia: A Community-Based Longitudinal Study," *The Lancet* 355, no. 9212 (2000): 1315–19.

36. Cacioppo, "What's Social."

37. Cacioppo and Patrick, *Loneliness*, 15.

38. Cacioppo and Patrick, *Loneliness*, 172.

39. Sheldon Cohen et al., "Social Ties and Susceptibility to the Common Cold," *JAMA* 277, no. 24 (1997): 1940–44.

40. Michael Marmot, *The Status Syndrome: How Social Standing Affects Our Health and Longevity* (Henry Holt, 2004), Introduction.

41. Author interview with Bert Uchino and Uchino lecture at University of British Columbia, October 30, 2014. For examples of relevant journal articles see Rebecca A. Campo et al., "The Assessment of Positivity and Negativity in Social Networks: The Reliability and Validity of the Social Relationships Index," *Journal of Community Psychology* 37, no. 4 (2009): 471–86; Julianne Holt-Lunstad et al., "On the Importance of Relationship Quality: The Impact of Ambivalence in Friendships on Cardiovascular Functioning," *Annals of Behavioral Medicine* 33, no. 3 (2007): 278–90.

42. For more on how they measure ambivalence see Rebecca A. Campo et al., "The Assessment of Positivity."

43. Julie K. Desjardins, Jill Q. Klausner, and Russell D. Fernald, "Female Genomic Response to Mate Information," *Proceedings of the National Academy of Sciences* 107, no. 49 (2010): 21176–80.

44. Lisa F. Berkman and Teresa Seeman, "The Influence of Social Relationships on Aging and the Development of Cardiovascular Disease: A Review," *Postgraduate Medical Journal* 62, no. 730 (1986): 805.

45. Except where noted, these details and all quotes are from author interview with Steve Cole.

46. Steve Cole speaking on "Social Regulation of Gene Expression" to the Foundation for Psychocultural Research, 2012.

47. Steve W. Cole, "Psychosocial Influences on HIV-1 Disease Progression: Neural, Endocrine, and Virologic Mechanisms," *Psychosomatic Medicine* 70, no. 5 (2008): 562–68.

48. Steve W. Cole et al., "Social Regulation of Gene Expression in Human Leukocytes," *Genome Biology* 8, no. 9 (2007): R189.

Chapter Four: Middle School Is about Lunch

1. Primary sources for this section are Konner, *Evolution of Childhood*, chap. 11; Willard W. Hartup and Nan Stevens, "Friendships and Adaptation in the Life Course," *Psychological Bulletin* 121, no. 3 (1997): 355; Willard W. Hartup, "Social Relationships and Their Developmental Significance," *American Psychologist* 44, no. 2 (1989): 120.

2. The concept of horizontal and vertical relationships is in Hartup, "Social Relationships."

3. "ask a preschooler . . .": Beverley Fehr, *Friendship Processes* (Sage, 1996), 16.

4. Another major source for this chapter is Kenneth H. Rubin et al., "Peer Interactions, Relationships, and Groups," in *Child and Adolescent Development: An Advanced Course*, eds. William Damon et al. (Wiley, 2008): 141–80.

5. Jerome Kagan, "Temperament," *Encyclopedia on Early Childhood Development* [online], (Montreal, Quebec: Centre of Excellence for Early Childhood Development, 2005): 1–4, Rev. April 2012.

6. Rubin et al., "Peer Interactions," 56.

7. Stuart Brown, *Play: How It Shapes the Brain, Opens the Imagination, and Invigorates the Soul* (Penguin Group, 2009), 5; see also Konner, *Evolution of Childhood*, 95.

8. Quoted in Emily Langer, "Jaak Panksepp, 'Rat Tickler' Who Revealed Emotional Lives of Animals, Dies at 73," *Washington Post*, April 21, 2001.

9. Brown, *Play*, 28.

10. Brown, *Play*, 32.

11. Gordon M. Burghart, "Defining and Recognizing Play," in *The Oxford Handbook of the Development of Play*, ed. Anthony D. Pellegrini (Oxford University Press, 2011); Sandra Aamodt and Sam Wang, "Part Four: The Serious Business of Play" in *Welcome to Your Child's Brain* (Oneworld, 2011).

12. Pamela Weintraub, "Discover Interview: Jaak Panksepp Pinned Down Humanity's 7 Primal Emotions," *Discover*, May 31, 2012.

13. Quoted in Brown, *Play*, 39.

14. Described in Brown, *Play*, 33–34. See also Marc Bekoff and John A. Byers, eds., *Animal Play: Evolutionary, Comparative and Ecological Perspectives* (Cambridge University Press, 1998).

15. Primary sources for this section are Konner, *Evolution of Childhood*; Rubin et al., *Peer Interactions*; and author interviews with Jaana Juvonen.

16. *CBS Evening News with Jeff Flor*, November 24, 2017.

17. Author interviews with Jaana Juvonen. Examples of papers that have resulted from this study are Hannah L. Schacter and Jaana Juvonen, "Dynamic Changes in Peer Victimization and Adjustment across Middle School: Does Friends' Victimization Alleviate Distress?" *Child Development* (2018); Leah M. Lessard and Jaana Juvonen, "Friendless Adolescents: Do Perceptions of Social Threat Account for Their Internalizing Difficulties and Continued Friendlessness?" *Journal of Research on Adolescence* 28, no. 2 (2018): 277–83.

18. Lessard and Juvonen, "Friendless Adolescents."

19. For a summary, see Jaana Juvonen, Guadalupe Espinoza, and Hannah Schacter, "Bullying," in *Encyclopedia of Mental Health*, 2nd ed. (Academic Press, 2018).

20. Schacter and Juvonen, "Dynamic Changes."

21. B. Bradford Brown, "The Role of Peer Groups in Adolescents' Adjustment to Secondary School," in *Peer Relationships in Child Development*, eds. T. J. Berndt and G. W. Ladd (Oxford, UK: John Wiley & Sons, 1989), 188–215.

22. David H. Hubel and Torsten N. Wiesel, "Receptive Fields, Binocular Interaction and Functional Architecture in the Cat's Visual Cortex," *Journal of Physiology* 160, no. 1 (1962): 106–54. For more of the story of this work see Lydia Denworth, *I Can Hear You Whisper: An Intimate Journey through the Science of Sound and Language* (Dutton, 2014), 31–33.

23. "Five Numbers to Remember about Early Childhood Development," Center on the Developing Child, Harvard University, www.developingchild.harvard.edu.

24. Jay N. Giedd et al., "Brain Development during Childhood and Adolescence: A Longitudinal MRI Study," *Nature Neuroscience* 2, no. 10 (1999): 861.

25. Interview with Jay Giedd for "Inside the Teenage Brain," *Frontline*, January 2002.

26. National Vital Statistics Report, vol. 67, no. 5, July 26, 2018; B. J. Casey, Rebecca M. Jones, and Todd A. Hare, "The Adolescent Brain," *Annals of the New York Academy of Sciences* 1124, no. 1 (2008): 111–26.

27. Casey et al., "The Adolescent Brain."

28. B. J. Casey, "Twelfth Jeffrey Lecture on Cognitive Neuroscience," UCLA, January 26, 2012.

29. Ben's story and its effect on Steinberg is included in Laurence Steinberg, *Age of*

Opportunity: Lessons from the New Science of Adolescence (Houghton Mifflin Harcourt, 2014), chap. 5.

30. These facts and figures about adolescent risk taking come from Steinberg, *Age of Opportunity*, 93.

31. Author interview with Laurence Steinberg. See also Dustin Albert, Jason Chein, and Laurence Steinberg, "The Teenage Brain: Peer Influences on Adolescent Decision Making," *Current Directions in Psychological Science* 22, no. 2 (2013): 114–20.

32. Sheree Logue et al., "Adolescent Mice, Unlike Adults, Consume More Alcohol in the Presence of Peers Than Alone," *Developmental Science* 17, no. 1 (2014): 79–85.

33. Steinberg, *Age of Opportunity*, 95.

34. Faustino et al., "Mechanisms of Social Buffering."

35. Wittig et al., "Social Support," 13361.

36. Leslie J. Seltzer, Toni E. Ziegler, and Seth D. Pollak, "Social Vocalizations Can Release Oxytocin in Humans," *Proceedings of the Royal Society B: Biological Sciences* 277, no. 1694 (2010): 2661–66.

37. Dylan G. Gee et al., "Maternal Buffering of Human Amygdala-Prefrontal Circuitry during Childhood but Not during Adolescence," *Psychological Science* 25, no. 11 (2014): 2067–78.

38. Ryan E. Adams, Jonathan Bruce Santo, and William M. Bukowski, "The Presence of a Best Friend Buffers the Effects of Negative Experiences," *Developmental Psychology* 47, no. 6 (2011): 1786.

39. Author interview with Megan Gunnar. This study and many others reported in Megan R. Gunnar, "Social Buffering of Stress in Development: A Career Perspective," *Perspectives on Psychological Science* 12, no. 3 (2017): 355–73.

Chapter Five: A Deep Wish for Friendship

1. I visited Amboseli with Alberts in March 2018.

2. Joan B. Silk, Susan C. Alberts, and Jeanne Altmann, "Social Bonds of Female Baboons Enhance Infant Survival," *Science* 302, no. 5648 (2003): 1231–34.

3. Nicholas K. Humphrey, "The Social Function of Intellect," in *Growing Points in Ethology*, eds. P. P. G. Bateson and R. A. Hinde (Cambridge University Press, 1976), 303–17.

4. Richard W. Byrne and Andrew Whiten, eds. *Machiavellian Intelligence: Social Expertise and the Evolution of Intellect in Monkeys, Apes, and Humans* (Oxford University Press, 1990).

5. Robin I. M. Dunbar, "The Social Brain Hypothesis," *Evolutionary Anthropology: Issues, News, and Reviews* 6, no. 5 (1998): 178–90.

6. Alex R. DeCasien, Scott A. Williams, and James P. Higham, "Primate Brain Size Is Predicted by Diet but Not Sociality," *Nature Ecology and Evolution* 1, no. 5 (2017): 0112.

7. Chris Venditti, "Evolution: Eating Away at the Social Brain," *Nature Ecology and Evolution* 1, no. 5 (2017): 0122.

8. Author interview with Robert Seyfarth.

9. Quotes from author interview with Mark Leary. Details also drawn from video interviews with Leary (https://vimeo.com/34785200) and the original paper: Roy F. Baumeister and Mark R. Leary, "The Need to Belong: Desire for Interpersonal Attachments as a Fundamental Human Motivation," *Psychological Bulletin* 117, no. 3 (1995): 497.

10. Baumeister and Leary, "Need to Belong."

11. Shelley E. Taylor, *The Tending Instinct: How Nurturing Is Essential to Who We Are and How We Live* (Macmillan, 2002), chap. 2, loc. 290, Kindle.

12. Shelley E. Taylor et al., "Biobehavioral Responses to Stress in Females: Tend-and-Befriend, Not Fight-or-Flight," *Psychological Review* 107, no. 3 (2000): 411.

13. Taylor, *Tending Instinct*, chap. 1, loc. 153–165.

14. Author interview with Jeanne Altmann; Jeanne Altmann, "Motherhood, Methods and Monkeys: An Intertwined Professional and Personal Life," in *Leaders in Animal Behavior: The Second Generation*, eds. Lee Drickamer and Donald Dewsbury (Cambridge University Press, 2009), 39–57.

15. Jeanne Altmann, "Observational Study of Behavior: Sampling Methods," *Behaviour* 49, no. 3–4 (1974): 227–66.

16. For the history of the Amboseli Project, see Susan C. Alberts and Jeanne Altmann, "The Amboseli Baboon Research Project: 40 Years of Continuity and Change," in *Long-Term Field Studies of Primates*, eds. Peter M. Kappeler and David P. Watts (Springer, Berlin: Heidelberg, 2012), 261–87.

17. Author interviews with Robert Seyfarth and Dorothy Cheney.

18. Dorothy L. Cheney and Robert M. Seyfarth, *Baboon Metaphysics: The Evolution of a Social Mind* (University of Chicago Press, 2008), 10–11.

19. For more on playbacks, see Dorothy L. Cheney and Robert M. Seyfarth, *How Monkeys See the World: Inside the Mind of Another Species* (University of Chicago Press, 1990).

20. Cheney and Seyfarth, *Baboon Metaphysics*, 12.

21. Dorothy Cheney, Robert Seyfarth, and Barbara Smuts, "Social Relationships and Social Cognition in Nonhuman Primates," *Science* 234, no. 4782 (1986): 1361–66.

22. Barbara B. Smuts, *Sex and Friendship in Baboons* (Aldine de Gruyter, 1985).

23. The story of Sylvia and Sierra is in Cheney and Seyfarth, *Baboon Metaphysics*, 87–88.

24. Author interview with Joan Silk.

25. Silk, "Using the 'F'-Word," 421.

26. Robin Sloan, *Mr. Penumbra's 24-Hour Bookstore* (Farrar, Straus and Giroux, 2012), loc. 448, Kindle.

27. Details on the paper about the strength of social bonds come from author interviews with Joan Silk and Susan Alberts.

28. See Silk, Alberts, and Altmann, "Social Bonds of Female Baboons," 1231–34; Thore J. Bergman et al., "Hierarchical Classification by Rank and Kinship in Baboons," *Science* 302, no. 5648 (2003): 1234–36.

29. Joan B. Silk, Jeanne Altmann, and Susan C. Alberts, "Social Relationships among Adult Female Baboons (*Papio cynocephalus*) I: Variation in the Strength of Social Bonds," *Behavioral Ecology and Sociobiology* 61, no. 2 (2006): 183–95; Joan B. Silk, Susan C. Alberts, and Jeanne Altmann, "Social Relationships among Adult Female Baboons (*Papio cynocephalus*) II: Variation in the Quality and Stability of Social Bonds," *Behavioral Ecology and Sociobiology* 61, no. 2 (2006): 197–204.

30. Joan B. Silk et al., "The Benefits of Social Capital: Close Social Bonds among Female Baboons Enhance Offspring Survival," *Proceedings of the Royal Society B: Biological Sciences* 276, no. 1670 (2009): 3099–104.

31. Joan B. Silk et al., "Strong and Consistent Social Bonds Enhance the Longevity of Female Baboons," *Current Biology* 20, no. 15 (2010): 1359–61.

32. Elizabeth A. Archie et al., "Social Affiliation Matters: Both Same-Sex and Opposite-Sex Relationships Predict Survival in Wild Female Baboons," *Proceedings of the Royal Society B: Biological Sciences* 281, no. 1793 (2014): 20141261.

33. Robert M. Seyfarth and Dorothy L. Cheney, "The Evolutionary Origins of Friendship," *Annual Review of Psychology* 63 (2012): 153–77.

34. Cacioppo, "What's Social."

Chapter Six: The Circles of Friendship

1. Daniel J. Hruschka, *Friendship: Development, Ecology, and Evolution of a Relationship*, vol. 5 (University of California Press, 2010).

2. Ethan Watters, *Urban Tribes: A Generation Redefines Friendship, Family, and Commitment* (Bloomsbury, 2003).

3. Tamas David-Barrett et al., "Communication with Family and Friends across the Life Course," *PloS One* 11, no. 11 (2016): e0165687.

4. Fehr, *Friendship Processes*, 82–83.

5. Michael Laakasuo et al., "The Company You Keep: Personality and Friendship Characteristics," *Social Psychological and Personality Science* 8, no. 1 (2017): 66–73.

6. All examples taken from Sarah H. Matthews, "Friendship Styles" in *Aging and Everyday Life*, eds. Jaber F. Gubrium and James A. Holstein (Blackwell, 2000), 155–94.

7. Martina Miche, Oliver Huxhold, and Nan L. Stevens, "A Latent Class Analysis of Friendship Network Types and Their Predictors in the Second Half of Life," *Journals of Gerontology Series B: Psychological Sciences and Social Sciences* 68, no. 4 (2013): 644–52.

8. Joseph Henrich, Steven J. Heine, and Ara Norenzayan, "The Weirdest People in the World?" *Behavioral and Brain Sciences* 33, no. 2–3 (2010): 61–83.

9. All of these examples taken from Hruschka, *Friendship*.

10. Hruschka, *Friendship*, 168.

11. Hruschka, *Friendship*, 184.

12. Author interview with Daniel Hruschka.

13. Robert L. Kahn and Toni C. Antonucci, "Convoys over the Life-Course: Attachment, Roles and Social Support," in *Life-Span Development and Behavior*, eds. Paul B. Baltes and Orville G. Brim (Academic Press, 1980), 253–86.

14. Author interview with Toni Antonucci.

15. For a detailed discussion of these statistics see Claude S. Fischer, *Still Connected: Family and Friends in America Since 1970* (Russell Sage Foundation, 2011), 44–49; Nicholas A. Christakis and James H. Fowler, *Connected: The Surprising Power of Our Social Networks and How They Shape Our Lives* (Little, Brown, 2009), 18.

16. Christakis and Fowler, *Connected*, 18.

17. The paper is Miller McPherson, Lynn Smith-Lovin, and Matthew E. Brashears, "Social Isolation in America: Changes in Core Discussion Networks over Two Decades," *American Sociological Review* 71, no. 3 (2006): 353–75. For details on the subsequent headlines see Fischer, *Still Connected*, 1.

18. For details see Fischer, *Still Connected*, 1–2; Hruschka, Santa Fe Institute lecture, July 20, 2012.

19. Keith N. Hampton, Lauren F. Sessions, and Eun Ja Her, "Core Networks, Social Isolation, and New Media: How Internet and Mobile Phone Use Is Related to Network Size and Diversity," *Information, Communication and Society* 14, no. 1 (2011): 130–55.

20. Christakis and Fowler, *Connected*, 18.

21. Fischer, *Still Connected*, 48.

22. Christakis and Fowler, *Connected*, 18; Toni C. Antonucci and Hiroko Akiyama, "Social Networks in Adult Life and a Preliminary Examination of the Convoy Model," *Journal of Gerontology* 42, no. 5 (1987): 519–27. Also author interview with Toni Antonucci.

23. Toni C. Antonucci, Kristine J. Ajrouch, and Kira S. Birditt, "The Convoy Model: Explaining Social Relations from a Multidisciplinary Perspective," *The Gerontologist* 54, no. 1 (2013): 82–92.

24. Author interview with Bert Uchino, Janice K. Kiecolt-Glaser, and Tamara L. Newton, "Marriage and Health: His and Hers," *Psychological Bulletin* 127, no. 4 (2001): 472–503.

25. Jacksonville versus Mexico City: H. Russell Bernard, Eugene C. Johnsen, Peter D. Killworth, Christopher McCarty, Gene A. Shelley, and Scott Robinson, "Comparing Four Different Methods for Measuring Personal Social Networks," *Social Networks* 12, no. 3 (1990): 179–215.

26. Eli J. Finkel, "The All-or-Nothing Marriage," *New York Times*, February 14. 2014, SR1.

27. A 1965 survey: Ethan Watters, "The Way We Live Now: In My Tribe," *New York Times Magazine*, October 14, 2001.

28. What people are looking for in a spouse: "Record Share of Americans Have Never Married," Pew Research Center, September 24, 2014.

29. 90 percent of us: Anna Goldfarb, "How to Maintain Sibling Relationships," *New York Times*, May 8, 2018.

30. Some 2000 Britons: Catherine Saint Louis, "Debunking Myths about Estrangement," *New York Times*, December 20, 2017.

31. In the United States . . .: Kylie Agllias, "Family Estrangement: Aberration or Common Occurrence?" *Psychology Today* (blog), September 8, 2014.

32. Social signatures: The first paper is Jari Saramäki et al., "Persistence of Social Signatures in Human Communication," *Proceedings of the National Academy of Sciences* 111, no. 3 (2014): 942–47. Larger studies are described in Sara Heydari et al., "Multichannel Social Signatures and Persistent Features of Ego Networks," *Applied Network Science* 3, no. 1 (2018): 8.

33. George A. Lundberg and Mary Steele, "Social Attraction-Patterns in a Village," *Sociometry* (1938): 375–419.

34. Christakis and Fowler, *Connected*, xiii–xiv. And author interviews with Nicholas Christakis.

35. Christakis and Fowler, *Connected*, 17.

36. Christakis and Fowler, *Connected*, 22.

37. The happiness study is described in Christakis and Fowler, *Connected*, 49–54; Clive Thompson, "Are Your Friends Making You Fat?" *New York Times Magazine*, September 10, 2009. Also the source of the Fowler quote.

38. Original work described in Mark S. Granovetter, "The Strength of Weak Ties," in *Social Networks*, ed. Samuel Leinhardt (Academic Press, 1977), 347–67.

39. Mario Luis Small, *Someone to Talk To* (Oxford University Press, 2017).

40. Quotes from author interview with Mario Luis Small.

41. Hruschka, *Friendship*, 140.

42. Jane Fonda and Lily Tomlin, TEDWomen, May 2015.

43. Hruschka, *Friendship*, 143.

44. Author interview with Jeffrey Hall.

45. Jeffrey A. Hall, "Sex Differences in Friendship Expectations: A Meta-Analysis," *Journal of Social and Personal Relationships* 28, no. 6 (2011): 723–47.

46. Author interview with Wendi Gardner. See also Elaine O. Cheung et al.,

"Emotionships: Examining People's Emotion-Regulation Relationships and Their Consequences for Well-Being," *Social Psychological and Personality Science* 6, no. 4 (2015): 407–14.

47. Arthur Aron et al., "The Experimental Generation of Interpersonal Closeness: A Procedure and Some Preliminary Findings," *Personality and Social Psychology Bulletin* 23, no. 4 (1997): 363–77.

48. See Daniel Jones, "Modern Love," *New York Times*, January 9, 2015.

49. Author interview with Beverley Fehr.

50. Fehr, *Friendship Processes*, 10–11.

51. Jeffrey A. Hall, "How Many Hours Does It Take to Make a Friend?," *Journal of Social Personal Relationships* 36, no. 4 (2019): 1278–96.

Chapter Seven: Digital Friendship

1. Details of Hollander's project are at www.areyoureallymyfriend.com.

2. Statistics from Pew Research Center: "Internet, Social Media Use and Device Ownership in U.S. Have Plateaued after Years of Growth," September 28, 2018; "Social Media Fact Sheet," February 5, 2018; "Teens, Social Media & Technology 2018," May 31, 2018.

3. Keith N. Hampton and Barry Wellman, "Lost and Saved . . . Again: The Moral Panic about the Loss of Community Takes Hold of Social Media," *Contemporary Sociology* 47, no. 6 (2018): 643–51.

4. Jean Twenge, "Have Smartphones Destroyed a Generation?" *The Atlantic*, September 2017.

5. Author interview with Jeff Hancock and Hancock's talk, "Psychological Well-Being and Social Media Use: A Meta-Analysis," at International Communication Association, May 2019.

6. Amy Orben and Andrew Przybylski, "The Association Between Adolescent Well-Being and Digital Technology Use," *Nature Human Behaviour* 3, no. 2 (2019): 173, and author interviews with Orben and Przybylski.

7. Author interviews with Ariel Shensa and Tracy Dennis-Tiwary.

8. Rebecca P. Yu et al., "The Relationships That Matter: Social Network Site Use and Social Wellbeing among Older Adults in the United States of America," *Ageing and Society* 36, no. 9 (2016): 1826–52.

9. Keith Hampton et al., "Social Networking Sites and Our Lives," *Pew Internet and American Life Project* 16 (2011): 1–85.

10. Hampton et al., "Social Networking," 42.

11. Brian A. Primack et al., "Social Media Use and Perceived Social Isolation among Young Adults in the US," *American Journal of Preventive Medicine* 53, no. 1 (2017): 1–8.

12. Author interview with Keith Hampton.

13. Jessica Vitak et al., "The Ties That Bond: Re-examining the Relationship between Facebook Use and Bonding Social Capital," in *2011 44th Hawaii International Conference on System Sciences* (IEEE, 2011), 1–10.

14. Sarah Myruski et al., "Digital Disruption? Maternal Mobile Device Use Is Related to Infant Social-Emotional Functioning," *Developmental Science* 21, no. 4 (2018): e12610.

15. Fischer, *Still Connected*, 9.

16. Fischer, *Still Connected*, 60.

17. Author interview with Candice Odgers.

18. Joanna C. Yau and Stephanie M. Reich, "Are the Qualities of Adolescents' Offline Friendships Present in Digital Interactions?," *Adolescent Research Review* 3, no. 3 (2018): 339–55.

19. Pew Research Center, "Teens, Technology & Friendships," August 2015.

20. Matt Richtel, "Growing Up Digital, Wired for Distraction," *New York Times*, November 21, 2010.

21. Amy Orben et. al., "Social Media's Enduring Effect on Adolescent Life Satisfaction," *Proceedings of the National Academy of Sciences* 116, no. 21 (2019): 10226–28.

22. Perri Klass, "When Social Media Is Really Problematic for Adolescents," *New York Times*, June 3, 2019.

23. Andrew K. Przybylski and Netta Weinstein, "A Large-Scale Test of the Goldilocks Hypothesis: Quantifying the Relations between Digital-Screen Use and the Mental Well-Being of Adolescents," *Psychological Science* 28, no. 2 (2017): 204–15.

Chapter Eight: Born to Be Friendly?

1. Author interviews with Nicholas Christakis.

2. See Carl Zimmer, *She Has Her Mother's Laugh: The Powers, Perversions and Potential of Heredity* (Dutton, 2018), 264–67. The book is an excellent source on the complexities of heredity, especially chap. 9.

3. James H. Fowler, Christopher T. Dawes, and Nicholas A. Christakis, "Model of Genetic Variation in Human Social Networks," *Proceedings of the National Academy of Sciences* 106, no. 6 (2009): 1720–24.

4. James H. Fowler, Jaime E. Settle, and Nicholas A. Christakis, "Correlated Genotypes in Friendship Networks," *Proceedings of the National Academy of Sciences* 108, no. 5 (2011): 1993–97.

5. Nicholas A. Christakis and James H. Fowler, "Friendship and Natural Selection," *Proceedings of the National Academy of Sciences* 111, no. 3 (2014): 10796–801.

6. Michael Platt in "Leaders in Social Neuroscience: An Interview with Dr. Michael Platt," posted September 3, 2013, on www.social-neuroscience.org (the site has since been reorganized and page no longer exists).

7. Author interview with Michael Platt.

8. For examples, see "Apes on a Plane" in Hrdy, *Mothers and Others*, 1.

9. For an example of what the Cayo social networks look like, see Lauren J. N. Brent et al., "Genetic Origins of Social Networks in Rhesus Macaques," *Scientific Reports* 3 (2013): 1042.

10. Lauren Brent, Lab Talk at the Human Nature Lab (headed by Nicholas Christakis), Yale University, November 5, 2015.

11. See, for example, Noah Snyder-Mackler et al., "Social Status Alters Chromatin Accessibility and the Gene Regulatory Response to Glucocorticoid Stimulation in Rhesus Macaques," *Proceedings of the National Academy of Sciences* 116, no. 4 (2019): 1219–28; Amanda J. Lea et al., "Dominance Rank-Associated Gene Expression Is Widespread, Sex-Specific, and a Precursor to High Social Status in Wild Male Baboons," *Proceedings of the National Academy of Sciences* 115, no. 52 (2018): E12163–71.

12. Zimmer, *She Has Her Mother's Laugh*, 277–81.

13. Brent, "Genetic Origins."

14. Personal details and direct quotes all from author interview with John Capitanio.

15. Author interview with John Capitanio. See also John P. Capitanio, "Individual Differences in Emotionality: Social Temperament and Health," *American Journal of Primatology* 73, no. 6 (2011): 507–15.

16. Author interview with Steve Cole.

17. Steven W. Cole et al., "Myeloid Differentiation Architecture of Leukocyte Transcriptome Dynamics in Perceived Social Isolation," *Proceedings of the National Academy of Sciences* 112, no. 49 (2015): 15142–47.

18. Cole, "Psychosocial Influences."

19. Erica K. Sloan et al., "Social Stress Enhances Sympathetic Innervation of Primate Lymph Nodes: Mechanisms and Implications for Viral Pathogenesis," *Journal of Neuroscience* 27, no. 33 (2007): 8857–65.

20. John P. Capitanio and Steven W. Cole, "Social Instability and Immunity in Rhesus Monkeys: The Role of the Sympathetic Nervous System," *Philosophical Transactions of the Royal Society B: Biological Sciences* 370, no. 1669 (2015): 20140104.

21. Unusually, this study includes both humans and macaques: Steven W. Cole et al., "Myeloid Differentiation Architecture of Leukocyte Transcriptome Dynamics in Perceived Social Isolation," *Proceedings of the National Academy of Sciences* 112, no. 49 (2015): 15142–47.

22. Capitanio and Cole, "Social Instability."

23. Gregory E. Miller et al., "A Functional Genomic Fingerprint of Chronic Stress in Humans: Blunted Glucocorticoid and Increased NF-κB Signaling," *Biological Psychiatry* 64, no. 4 (2008): 266–72.

24. For a summary, see Steven W. Cole, "Human Social Genomics," *PLoS Genetics* 10, no. 8 (2014): e1004601.

Chapter Nine: Deeply Built into the Brain

1. Daniel P. Kennedy and Ralph Adolphs, "The Social Brain in Psychiatric and Neurological Disorders," *Trends in Cognitive Sciences* 16, no. 11 (2012): 559–72.

2. Author interview with Thalia Wheatley.

3. See J. P. Mitchell and T. F. Heatherton, "Components of a Social Brain," *Cognitive Neurosciences IV* (2009): 951–58; Ralph Adolphs, "The Social Brain: Neural Basis of Social Knowledge," *Annual Review of Psychology* 60 (2009): 693–716.

4. I adapted my example of walking into a party from one presented in Sébastien Tremblay, K. M. Sharika, and Michael L. Platt, "Social Decision-Making and the Brain: A Comparative Perspective," *Trends in Cognitive Sciences* 21, no. 4 (2017): 265–76. The paper also provided details of relevant brain areas.

5. Edith Wharton, *A Backward Glance: Reminiscences* (New York: D. Appleton & Company, 1934), 115.

6. Arthur Aron and Barbara Fraley, "Relationship Closeness as Including Other in the Self: Cognitive Underpinnings and Measures," *Social Cognition* 17, no. 2 (1999): 140–60.

7. Wendi L. Gardner, Shira Gabriel, and Laura Hochschild, "When You and I Are 'We,' You Are Not Threatening: The Role of Self-Expansion in Social Comparison," *Journal of Personality and Social Psychology* 82, no. 2 (2002): 239.

8. These examples are discussed in more detail in Hruschka, *Friendship*, 30–34.

9. Tania Singer et al., "Empathy for Pain Involves the Affective but Not Sensory Components of Pain," *Science* 303, no. 5661 (2004): 1157–62.

10. Lydia Denworth, "I Feel Your Pain," *Scientific American* 317, no. 6 (2017): 58–63.

11. Lane Beckes, James A. Coan, and Karen Hasselmo, "Familiarity Promotes the Blurring of Self and Other in the Neural Representation of Threat," *Social Cognitive and Affective Neuroscience* 8, no. 6 (2012): 670–77.

12. See also Lauren J. N. Brent et al., "The Neuroethology of Friendship," *Annals of the New York Academy of Sciences* 1316, no. 1 (2014): 1–17.

13. Wei Song Ong, Seth Madlon-Kay, and Michael L. Platt, "Neuronal Mechanisms of Strategic Cooperation," *BioRxiv* (2018): 500850.

14. Much of this work is summarized in Tremblay et al., "Social Decision-Making."

15. Randi Hutter Epstein, *Aroused: The History of Hormones and How They Control Just about Everything* (W. W. Norton & Company, 2018), 7.

16. Paul J. Zak, *The Moral Molecule: How Trust Works* (Penguin, 2013), 23.

17. Jennifer N. Ferguson et al., "Oxytocin in the Medial Amygdala Is Essential for Social Recognition in the Mouse," *Journal of Neuroscience* 21, no. 20 (2001): 8278–85; Zoe R. Donaldson and Larry J. Young, "Oxytocin, Vasopressin, and the Neurogenetics of Sociality," *Science* 322, no. 5903 (2008): 900–4.

18. Steve W. Chang and Michael L. Platt, "Oxytocin and Social Cognition in Rhesus

Macaques: Implications for Understanding and Treating Human Psychopathology," *Brain Research* 1580 (2014): 57–68.

19. For a more extended discussion, see Tremblay et al., "Social Decision-Making."

20. Catherine Crockford et al., "Urinary Oxytocin and Social Bonding in Related and Unrelated Wild Chimpanzees," *Proceedings of the Royal Society B: Biological Sciences* 280, no. 1755 (2013): 20122765.

21. James P. Burkett et al., "Oxytocin-Dependent Consolation Behavior in Rodents," *Science* 351, no. 6271 (2016): 375–78.

22. Author interview with James Burkett.

23. Author interview with Jamil Zaki.

24. Sandra Manninen et al., "Social Laughter Triggers Endogenous Opioid Release in Humans," *Journal of Neuroscience* 37, no. 25 (2017): 6125–31.

25. Author interviews with Carolyn Parkinson.

26. Carolyn Parkinson, Adam M. Kleinbaum, and Thalia Wheatley, "Similar Neural Responses Predict Friendship," *Nature Communications* 9, no. 1 (2018): 332.

27. Described to me by Thalia Wheatley. At press time, the study was still unpublished.

28. Author interview with Thalia Wheatley and Emma Templeton.

29. Dana Bevilacqua et al., "Brain-to-Brain Synchrony and Learning Outcomes Vary by Student–Teacher Dynamics: Evidence from a Real-World Classroom Electroencephalography Study," *Journal of Cognitive Neuroscience* 31, no. 3 (2019): 401–11; Kelong Lu and Ning Hao, "When Do We Fall in Neural Synchrony with Others?" *Social Cognitive and Affective Neuroscience* (2019).

30. For a review, see Uri Hasson and Chris D. Frith, "Mirroring and Beyond: Coupled Dynamics as a Generalized Framework for Modelling Social Interactions," *Philosophical Transactions of the Royal Society B: Biological Sciences* 371, no. 1693 (2016): 20150366.

31. Joy Hirsch et al., "Frontal Temporal and Parietal Systems Synchronize within and across Brains during Live Eye-to-Eye Contact," *Neuroimage* 157 (2017): 314–30. And author interview with Joy Hirsch.

Chapter Ten: The Good Life, Revealed

1. Details of Paula Dutton's experience taken from both author interview with Dutton and an account by Veronique de Turenne, "Lonely Planet," *U Magazine*, UCLA Health, Fall 2016.

2. Details of the Generation Xchange program from author interviews with Teresa Seeman and codirector D'Ann Morris.

3. Berkman and Seeman, "Influence of Social Relationships."

4. Teresa E. Seeman et al., "Social Network Ties and Mortality among the Elderly in the Alameda County Study," *American Journal of Epidemiology* 126, no. 4 (1987): 714–23.

5. Author interviews with Paula Dutton, Linda Ricks, Michael Travers, and Theresa Brissett.

6. Jo Marchant, "Immunology: The Pursuit of Happiness," *Nature News* 503, no. 7477 (2013): 458.

7. Lucy Rock, "Life Gets Better after 50: Why Age Tends to Work in Favour of Happiness," *The Guardian*, May 5, 2018.

8. David G. Blanchflower and Andrew J. Oswald, "Is Well-Being U-Shaped over the Life Cycle?" *Social Science and Medicine* 66, no. 8 (2008): 1733–49.

9. Laura L. Carstensen, Helene H. Fung, and Susan T. Charles, "Socioemotional Selectivity Theory and the Regulation of Emotion in the Second Half of Life," *Motivation and Emotion* 27, no. 2 (2003): 103–23.

10. All statistics from the Pew Research Center.

11. *The Loneliness Project* video published to YouTube on September 22, 2017, by Campaign to End Loneliness.

12. "Loneliness and Social Isolation in the United States, the United Kingdom, and Japan: An International Survey," The Kaiser Family Foundation, 2018; Cigna Survey, 2018; "Loneliness and Social Connections," AARP Survey, 2018.

13. Yang Claire Yang et al., "Social Relationships and Physiological Determinants of Longevity across the Human Life Span," *Proceedings of the National Academy of Sciences* 113, no. 3 (2016): 578–83.

14. William J. Chopik, "Associations among Relational Values, Support, Health, and Well-Being across the Adult Lifespan," *Personal Relationships* 24, no. 2 (2017): 408–22.

15. Author interview with Julianne Holt-Lunstad.

16. Details reported in Katie Hafner, "Researchers Confront an Epidemic of Loneliness," *New York Times*, September 5, 2016.

17. Author interview with Lisa Berkman.

18. Susan Pinker, *The Village Effect: How Face-to-Face Contact Can Make Us Healthier and Happier* (Spiegel & Grau, 2014), 51, 54.

19. Pinker, *Village Effect*, 54.

20. Julianne Holt-Lunstad, Theodore F. Robles, and David A. Sbarra, "Advancing Social Connection as a Public Health Priority in the United States," *American Psychologist* 72, no. 6 (2017): 517.

21. Joan DelFattore, "If You're Single with Cancer, You May Get Less Aggressive Treatment Than a Married Person," *Washington Post*, December 3, 2018.

22. Janet Adamy and Paul Overberg, "The Loneliest Generation: Americans, More Than Ever, Are Aging Alone," *Wall Street Journal*, December 11, 2018.

23. Details from George E. Vaillant and Kenneth Mukamal, "Successful Aging," *American Journal of Psychiatry* 158, no. 6 (2001): 839–47; Joshua Wolf Shenk, "What Makes Us Happy?" *The Atlantic*, June 2009.

24. Robert Wadlinger, "What Makes a Good Life? Lessons from the Longest Study on Happiness," TedXBeaconStreet, November 2015.

Index

5-HTTLPR, 196
"36 Questions That Lead to Love," 158–59

AARP, 77
A-beta fibers, 57–58
Abrams, Daniel, 56
acceptance. *See* social acceptance
Acid (baboon), 117, 137
acquaintances, 161, 171–72, 173
acquisitive friendship style, 143, 183
Adams, Geoff, 215
adolescence, 91–115, 177. *See also*
　　teenagers
　emotional reactivity and, 106
　friendlessness and, 101
　frontal cortex and, 105–6
　imbalance model of adolescent brain
　　development, 106
　"peer effect" and, 107–10
　peer presence and, 108–11
　peer relationships and, 98–99
　"social brain" and, 110–11
　stress and, 113–14
adrenaline, 68–69, 96
adult brains, social stimuli and, 207–8

adults
　digital technology and, 174
　living alone, 241–43 (*see also* singledom;
　　widows/widowers)
　older adults, 169–70, 241–42 (*see also*
　　aging)
　social media and, 169–70
adversity, 205
Aegyptopithecus, 199
affection, 25
affective touch, 57–59
African Americans, 101
agency, 157
"age of reason," 92–93
aggression/aggressiveness, 2, 11–13, 41, 42,
　　80, 96, 119, 196
aging, 231–32
　cellular aging, 84, 86
　happiness and, 239–40
　health and, 241–42
　loneliness and, 231–32, 241–42
　primates and, 240–41
　social networks and, 239–41
　social relationships and, 87–88, 232
　stress and, 87–88

agreeableness, 94, 142
AIDS, 88–89
Alameda County, California, data from,
    67–68, 71, 72, 87, 232–33
Alberts, Susan, 116, 133–37, 246
Alex, 91–92, 114, 177–78
"all-or-nothing marriage," 149
Altmann, Jeanne, 125–26, 128, 133–34,
    135, 192
Altmann, Stuart, 31–32, 34–39, 125–26, 132
altruism, 34, 39, 40
ambivalence, relationships and, 85–86
Amboseli Baboon Research Project,
    116–17, 125–27, 132, 133–37, 194,
    200, 246
Amboseli National Park, 116. *See also*
    Amboseli Baboon Research Project
American Sociological Review, 147
Amy, 156, 157
amygdala, 47, 113, 214
animal nature, human nature and, 37–38
animal research. *See also specific projects*
    comparative studies across species,
        199–200
    human research and, 126–27
    loneliness and, 200–204
    Machiavellian intelligence hypothesis,
        119–20
    social network analysis and, 191–95
    stress and, 68, 111–12
animals, 136. *See also* animal research;
    specific animals
    playing and, 94–95
    social behavior of, 6–7
    social bonds and, 87, 117–18
Ann, 184
"anomie," 25
anterior cingulate cortex (ACC), 209, 211,
    214, 219
anterior insula, 211

anthropodenial, 7
anthropologists, 118, 139
anthropomorphism, 7, 131
Antonucci, Toni, 146–47
anxiety, 101, 166, 168, 169, 179, 229–30,
    236–37
apes, 119, 176
archeologists, 5
Archie, Elizabeth, 135
"Are You Really My Friend?," 164
Arianne, 139
Arizona State University, 131, 144
Aristotle, 24, 136, 156, 162
Aron, Arthur, 158–59, 209
atevi, 144
Atlantic, 170
attachment, 20–42, 45, 100, 174–75
attachment theory, evolutionary theory
    and, 30–31, 45
auditory cortex, 222–23
auditory cues, 49
Aurora, 45–46, 47, 56, 59
autism, 6, 59–60, 197, 218
autonomy, 83–84

Baboon Camp, 130–32
baboons, 15, 116–17, 120, 128, 130, 131,
    134, 136–37, 176, 194, 197, 200, 246
Babylab at the Centre for Brain and Cog-
    nitive Development at Birkbeck,
    University of London, 45–46, 48–49,
    56–57, 59
Bacon, Kevin, 151
Bangladesh, 146
Barbary macaques, 240–41
Barrett, Lisa Feldman, 51
Barry, 242–43
Baumeister, Roy, 121–24, 126, 162
behaviorism, 25
Belkaya, Naz, 214

belonging
   desire for, 121–24, 162
   as "master motive," 121–24
Bennington College, 151
Berkman, Lisa, 70–71, 72, 76, 87–88, 232,
      245, 246, 247
Bernice, Miss, 236
Berntson, Gary, 75–77
Berwick, Donald, 248
best friends, 93, 148–49, 161. See also close
      friendships
biochemistry, 217–28
biological determinism, 42
biology
   psychology and, 24–25, 87
   social behavior and, 190–92
birds, 28–29, 32, 44, 49
Birkbeck, University of London, 44. See
      also Babylab at the Centre for Brain
      and Cognitive Development at Birk-
      beck, University of London
blood pressure, 7, 79–80, 84, 85, 86, 87, 238
Blue Zones, 245–46
Blum, Deborah, 25–26, 30
Bock, Arlie, 249–50
"body channel," of empathy, 62
Boncz, Adam, 227–28
bonobos, 118
Botswana, 130–32
Bowlby, John, 20–22, 25–26, 29–31, 45,
      127, 146, 174–75
boys, 158. See also gender
   friendlessness and, 101
   video games and, 177–78
Brad, 206–7, 220–23, 224–25
brain cells, 4, 104, 105
brain imaging, 47–48, 75, 206–7, 211, 213,
      220–23, 224–25, 226
   children and, 104–5
   hyperscanning, 227–28

infants and, 48–49, 54–56
   limitations of, 227
   social network analysis and, 222–24
brain lesions, 216–17
brain processing, friendship and, 212, 221–22
brains, 4, 17, 39. See also neuroscience
   anatomy of, 46, 105 (see also specific
      anatomy)
   brain regions, 46–47, 207, 222–25
   deeply built into, 206–28
   development of, 44
   of infants, 47–48
   lobes of, 46–47
   maturation of, 105–6
   networks of, 46–47
   size of, 97, 119, 120
   social brain, 43–64, 110–11, 119–21,
      207, 208–9, 216–17
   teenage, 103–11
brain synchrony, 222–25, 226
Brent, Lauren, 2–4, 7, 11, 13–15, 34, 190–
      92, 194–97
Brian, 133
Brigham Young University, 244
Brissett, Theresa, 230–31, 235
Brown, Stuart, 94–95, 96
bullying, 100, 101–2, 177
Burkett, James, 218–19
Burns, George, 85
Byers, John, 97
Byrne, Richard, 119–20

Cacioppo, John, 74–82, 84, 88–90, 101,
      136, 200–201, 203–5, 231–32, 244
Cajal (rhesus macaque), 213–14
Caldwell, Gail, 222
California National Primate Research cen-
      ter, 198–200
Cambridge University. See University of
      Cambridge

Campaign to End Loneliness, 242–43

cancer, 67–68, 203, 205, 247–48

Capitanio, John, 198–200, 201, 202–4

Caraballo, Giselle, 3–4, 10, 11, 13

cardiovascular function, 79

caregivers, stress and, 76

Caribbean Primate Research Center, 1. *See also* Cayo Santiago

Carnegie Mellon University, 82

Carpenter, Clarence, 34–36

Carstensen, Laura, 240

Case Western Reserve University, 122

Casey, B. J., 106

Cassel, John, 69–70

Catherine, 140

Catholic University, 23

Cayo Santiago, 1–4, 6, 9–11, 14–16, 32, 34–39, 128, 190–92, 196–97, 212, 216–17

cell phones, 165, 169, 171, 172, 178
  disconnection caused by, 174–75

cellular aging, 84, 86

Central European University, 227

centrality (social network analysis), 188, 189, 197

cerebellum, 97, 220

C fibers, 57

Chandrashekar, Aparna, 192–94

Charlie, 177

Chein, Jason, 109–10

chemistry, friendship and, 140–41

Cheney, Dorothy, 127–32, 134, 190, 213, 218, 240

Cherryh, C. J., 144

Cheung, Elaine, 158

chickens, 49–50

childhood
  length of human, 44, 97
  loneliness and, 91–92
  making friends and, 91–92
  research on, 25–42

childlessness, 241–42, 247–48

children, 92. *See also* infants; teenagers; toddlers
  brain development of, 47–48, 92–93, 103–11, 207
  brain imaging and, 104–5
  five-to-seven shift, 92–93
  friendships of, 17–19, 43–44, 63–64, 251–52
  sociability in, 59–60
  technology and, 179–80

chimpanzees, 32, 63, 111–12, 118, 119, 120, 136, 176, 218

cholesterol, 238

Chopik, William, 243

Christakis, Nicholas, 152–53, 185–90, 196–97, 224, 246

Christian, 17–19, 60–61, 63, 94, 179

chronic disease, 67–68, 203. *See also specific diseases*

Cigna, 77

Cleveland State University, 143

cliques, 14, 102, 103, 151–52

close friendships, 146–49, 150–51, 156–57, 170, 172, 173
  mapping of, 212
  proximity of, 246

closeness, 179, 209, 212. *See also* close friendships; proximity

Coan, James, 211–12

Cobb, Sidney, 69–70

cognitive empathy, 61, 62–63

cognitive science, 209

Cohen, Sheldon, 82–83

Cole, Steve, 88–89, 201–5, 238, 249, 250

collaboration, 93, 227

Columbia University, 34–35

communication, 18, 32, 47, 51–52, 118

communities, 69–70, 245–46, 247
  "circulatory system" of, 151
  health and, 67–68

companionship, 177

compassion, 61

confidants, 70, 147–48, 150, 154–55, 172

conflict resolution, 177

connection. *See* social connections

conscientiousness, 94, 142

conserved transcriptional response to
    adversity (CTRA), 205

consoling, 219

contagion, 151, 153

conversations, 162, 225–27

cooperation, 63, 93, 208

cooperativeness, 42, 63. *See also*
    cooperation

"core discussion network," 146

coronary artery calcification, 84, 86

corporations, 247

corpus callosum, 220

cortical brain regions, 207

cortisol, 112–13, 114

coworkers, 172

creation, 95

creative intellect, role in holding society
    together, 118–19

creativity, 42, 95, 118–19

Crete, 144

Crockford, Cathy, 218

crowds, 102

C-tactile (CT) afferents, 57–58, 59

cultural values, 121

cultural variation, in friendship, 143–46

Dan, 65, 70, 140

Darwin, Charles, 33–34, 39, 128, 131–32

data sets, 168, 232–33
    from Alameda County, California,
        67–68, 71, 72, 87–88, 232–33
    field data, 126
    from Framingham, Massachusetts,
        67–68, 153
    from Tecumseh, Michigan, 67–68, 72

Dawkins, Richard, 41, 42

decade of the brain, 74–75

degree (social network analysis), 188

De Jong Gierveld, Jenny, 78

DelFattore, Joan, 247–48

demographic change, 241–42

Dennis-Tiwary, Tracy, 169, 175

depression, 80, 85, 101, 166, 168, 169, 179,
    229–30

De Waal, Frans, 7, 61, 62, 119

Diamond, Marian, 96–97

diet, brain size and, 120

digital friendship, 163–82

digital technology, disruption caused by,
    174–75

direct fitness, 40

discerning friendship style, 143

disgust (emotion), 51

disliking, 41

distance (between friends), 212

diversity, 82–83, 103

DNA, 88, 205. *See also* genetics

dolphins, 6, 136, 246

dopamine, 96, 217

dopaminergic midbrain, 209

dorsolateral prefrontal cortex, 209

ducks, 28

Duke University, 116, 190

Dunbar, Robin, 119–20, 146

Durkheim, Émile, 25

Dutton, Paula, 229–31, 234, 235, 236–37,
    239

dying, fear of, 121

eating disorders, 179

education, 143, 147, 172

EEG, 226

eigenvector centrality, 188

Einfühlung, 61

EKG, 59

elephants, 136

Elizabeth, 156, 185

e-mail, 174

Emerson, Ralph Waldo, 23

Emory University, 61, 217, 218

emoticons, 177

emotion(s), 24, 38–39, 41, 58, 219

    communication of, 51–52

    emotional empathy, 61

    emotional energy, 162

    emotional nourishment, 157

    emotional reactivity, 106

    emotion contagion, 61–62

    faces and, 51–52

    measurement of, 74

    physiology and, 74

    regulation of, 208

    social network analysis and, 153

empathy, 61, 119, 211

    "body channel" of, 62

    cognitive, 61, 62–63

    emotional, 61

    empathic concern, 61, 62

    playing and, 94–95

    sense of self and, 61–62

    three main components of, 61

endocrine gland, 217

endorphins, 217, 220

"enriched environments," 96–97

entertaining, 183–84

epidemiology, 67–68, 83–84, 146, 246. See
    also specific research projects

Epstein, Randi Hutter, 217

Estimon, Denis, 99

ethnic diversity, 103

ethology, 27–28, 38, 39

eudemonic happiness, 169

Evans County, Georgia, 72

evolution, 15, 16, 17, 197

evolutionary biologists, 118

evolutionary biology, 116–17, 128–29

evolutionary science, 33

evolutionary theory, 33–34, 39, 40, 121, 129

    attachment theory and, 30–31

    friendship and, 132–33

extroversion, 94, 142

eye contact, 174–75

Facebook, 163–64, 165, 170, 171–72, 173,
    180–82

face perception, infants and, 52–55

faces, 50–52

    emotion and, 51–52

    face perception and, 52–55

    facial expressions, 51–55

    facial recognition, 53

    focusing on, 15–16

Fagen, Bob, 95

familiarity, 12, 24, 211–12

family friendly policies, 247

family members, 4, 85–86, 136, 149–50,
    173, 240

    extended, 172, 248

    immediate, 172

    proximity to, 246

fear (emotion), 39, 51, 204

fear of missing out (FOMO), 166

Fehr, Beverley, 159

"fellow feeling," 24

field data, 126

field observation, 37

fight-or-flight response, 124–25, 205

financial issues, social isolation and, 248

Finkel, Eli, 149

Fischer, Claude, 176

Fischer, Julia, 240

fish, 44, 87, 205

Fisher, Ronald, 33–34

fission-fusion societies, 136

fitness, 40

five-to-seven shift, 92

focal sampling, 126

Fonda, Jane, 157

foraging, 120

Fossey, Dian, 30

Fowler, James, 152–53, 186, 187–90, 196–97, 224

Framingham, Massachusetts, data from, 67–68, 153

Framingham Heart Study, 68, 153

Freud, Sigmund, 25, 92

Freudian psychology, 30, 31

friendlessness, adolescence and, 101

friendliness. *See also* sociability, genetics and, 183–205

friends, 4, 173. *See also* friendship(s); *specific kinds of friendship*
  calming effect of, 7
  from college, 14–15, 172, 181–82
  as "functional kin," 190
  from high school, 172, 181
  influence of, 65–66
  making, 91–115

friendship(s). *See also specific kinds of friendship*
  acquisitive style, 143, 183
  animal research on, 6–7, 136
  biological and evolutionary foundations of, 6, 7, 15, 16, 246
  biology and, 20–42. *See also* health
  brain processing and, 212, 221–22
  as a buffer, 102, 111–14
  casual, 161
  chemistry and, 140–41
  of children, 17–19, 43–44, 63–64
  circles of, 138–62
  close, 146–49, 150–51, 156–57, 170, 172, 173, 212, 246
  costs of, 141
  cultural variation in, 143–46, 158
  dearth of research on, 5
  deep wish for, 116–37
  definition of, 5, 29, 132–33
  digital, 163–82
  discerning style, 143
  evolutionary advantage of, 40
  evolutionary theory and, 132–33
  expectations of, 157–58
  failure to take seriously, 5–6
  falling away of, 141
  female vs. male, 156–57
  friendship styles, 183
  to fulfill different needs, 158
  groups, 158
  health and, 65–90
  independent style, 143
  mapping of, 212
  mismatched, 141
  natural selection and, 130, 134–35
  one-on-one, 158
  optimism about, 166–67
  prioritizing, 162
  as a priority, 248–49
  proximity of, 246
  quality of, 6, 84–86, 135, 250–51
  quantity of, 6, 250
  sensory nature of, 17
  under the skin, 65–90
  stress and, 111–12
  three distinct styles of, 143
  value of, 139, 157
  visible and invisible aspects of, 16

frontal cortex, 97, 105–6, 220

frontal lobe, 46

fruit flies, 213

functional magnetic resonance imaging (fMRI), 54, 82, 206–7, 211, 225, 227–29

functional near-infrared spectroscopy (fNIRS), 48–49, 54–56, 226

Gardner, Wendi, 158

Garner, Menora, 236

gay men, AIDS and, 88–89, 201

Gee, Dylan, 113–14

geese, 28–29

gender, 94, 124–25; 127, 156–57, 158–59
gene expression, 201, 203, 249
    immune system and, 238
    loneliness and, 89–90, 200–205
    social effects of, 88–89
General Social Survey, 147–48
Generation Xchange, 230–31, 233–39
genes, 17, 40, 42, 88–89. *See also* gene
        expression; genetics; genomes
    health and, 201
    hierarchy and, 195–96
    inflammation and, 204–5
    morality and, 37–38
    sociability and, 183–205
    social behavior, 38–39
    social behavior and, 207, 216–17
genetics, 24, 34, 183–205
Genette, 57–58, 59
genomes, 88, 89, 195–96, 216–17
    genome-wide association studies
        (GWAS), 195–96
    genomic analysis, 88–90, 238
genotype, 187, 189–90
geographic dislocation, 246, 247. *See also*
        mobility; proximity
Georgina, 45–46
German Primate Center, Cognitive Ethol-
        ogy Lab, 240–41
Giedd, Jay, 104–5
girls, 158, 178–79. *See also* gender
glucocorticoids, 68–69, 112–13
Goldilocks hypothesis, 180
Gombe Reserve, 32, 120, 127
Goodall, Jane, 20, 29–30, 32, 120, 127
the good life, 229–52. *See also* happiness;
        well-being
Goren, Carolyn, 52
gorillas, 30
Gould, Stephen Jay, 42
Granovetter, Mark, 153–54

gratitude, 41
gray matter, 105
Greek philosophers, 23–24
gregariousness, 91–92, 94, 183–205. *See*
        also sociability
grooming, 9, 12, 13, 16, 18, 116, 117, 130,
        135, 137, 176, 196, 197
groups, 102, 103, 120, 158
guilt, 39, 41
Gunnar, Megan, 114

habits, 65–66
Hall, Jeff, 157, 160–62
Hamilton, William, 39–40, 42
Hampton, Keith, 166, 170, 171, 173–74
Hancock, Jeff, 167, 169–70, 173
happiness, 153
    aging and, 239–40
    eudemonic happiness, 169
    hedonic happiness, 169
    measurement of, 166
happiness hormones, 217–20
Harlow, Harry, 30, 36, 198
Harvard University, 31–33, 42, 40, 42, 54,
        71, 127, 154, 227–28, 246, 249–50
Hasson, Uri, 226, 227
hate, 39
Hawkley, Louise, 78, 79–80, 101
health, 7, 73, 143, 148
    aging and, 241–42
    community and, 67–68
    friendship and, 65–90
    genes and, 201
    loneliness and, 80–82, 231–32, 241–42,
        248
    relationships and, 84–87
    social connectedness and, 76–78, 238,
        250–51
    social isolation and, 70–71, 74
    social networks and, 80

social relationships and, 65–90
social status and, 83–84
stress and, 201
health care system, 247–48
loneliness and, 248
marital status and, 248
hearing, 17, 55
heart disease, 67–68, 203
hedonic happiness, 169
helplines, 244
heredity, 33. *See also* genetics
heritability, 187–89
social networks and, 188–89
herpes simplex, 75–76
hierarchy, 9–11, 13, 36, 101–2, 103
baboons and, 131
genetics and, 195–96
rhesus macaques, 9–11, 13
Hinde, Robert, 20–22, 29–30, 31, 127, 140
Hirsch, Joy, 226
HIV genome, 89
HIV virus, 201
stress and, 202–3
Hobbes, Thomas, 166
Hollander, Tanja, 163–64, 180
Holt-Lunstad, Julianne, 74, 244–45, 246–47
homology, 199–200
horizontal relationships, 93
hormones, 68–69, 112–13, 114, 191–92, 196, 217–18, 220. *See also specific hormones*
House, James, 69–70, 71–74, 76, 146, 173
Hrdy, Sarah Blaffer, 43
Hruschka, Daniel, 144, 145, 146
Hubel, David, 103–4
hugging, 241
human nature, animal nature and, 37–38
human research, animal research and, 7–8, 126–27

Humphrey, Nicholas, 118–19
Hurricane Maria, 1
hyenas, 136
hyperscanning, 227–28
hypothalamic pituitary adrenal (HPA) axis, 112
hypothalamus, 39, 114

immune cells, 7
immune system, 85, 87, 204, 238, 246
immunology, 89–90
stress and, 75–76
imprinting, 28–29, 49–50
inclusive fitness, 40
independent friendship style, 143
indirect fitness, 40
individuality, 143
individuals, 102–3
infants
affective touch and, 59
brain development of, 47–48, 207
brain imaging and, 48–49, 54–56
face perception and, 52–55
facial expressions and, 52–55
highly reactive, 94
highly social, 94
mutual gaze and, 59
play and, 97–98
social smile and, 59
touch and, 56–59
voice recognition and, 55–56
inflammation, 84, 86, 202–3, 204
genes and, 204–5
loneliness and, 89–90
information, exchanges of, 154
inhibition, 93–94
injustice, 41
inner circles, 146–49, 150–51, 171, 172
insects, social, 39
instability, 100

Instagram, 165, 177
Institute for Social Research, University of
    Michigan, 71–72
institutions, 247
insular cortex, 58
interest (emotion), 51
interference effect, 82
internet, 165, 172
intimacy, 93, 100, 161
    patterns of, 150
    social media and, 172–73
    styles of, 157
introversion, 93–94, 200. See also shyness
isolation. See social isolation
Ivy (baboon), 117, 137

Jacob (Jake), 17–19, 43–44, 60–61, 63,
    91–92, 94, 98, 138, 141, 179, 187,
    251–52
James, William, 24–25, 74
Jason, 115
Jefferson, Thomas, 166
Jenny, 154
Jiang, Yaoguang, 213–14, 216
Joe, 242–43
John D. and Catherine T. MacArthur
    Foundation, 76–77, 88
Johnson, Mark, 44, 49–50, 51–53, 54, 55,
    59, 226
joy (emotion), 51
Julie, 154
justice, 42
Juvonen, Jaana, 99–103

Kagan, Jerome, 93–94
Kahn, Robert, 146
Kanwisher, Nancy, 45–46, 53, 54
Kat, 154
Kathryn (baboon), 137
Kenya, 116, 125–26, 130, 132

Kiecolt-Glaser, Janice, 76, 84
kinship, 39, 136. See also family members
Kipling, Rudyard, 129
Kleinbaum, Adam, 221–23
Knapp, Caroline, 222

Landis, Karl, 72–73
language, 92, 118. See also
    communication
Latinos, 101
laughter, 8–9, 18
Leah, 181
Leakey (rhesus macaque), 213–14
Leakey, Louis, 30, 120
Leary, Mark, 121–24, 126, 162
Lewontin, Richard, 42
Lepcha farmers, 144
Lessard, Leah, 101
leukocytes, 89–90
Lewis, C. S., 5, 6
life satisfaction, 169, 178
life span, relationships across the, 141–42
liking, 41
limbic system, 39, 106, 107, 207
Liverpool John Moores University, 58
Livingston, Margaret, 54
Lloyd-Fox, Sarah, 48, 49, 54–56, 60,
    97–98, 226
loneliness, 16–17, 136, 166, 169, 173, 203,
    205, 229–30
    as adaptive response, 81–82
    adolescence and, 91–92
    aging and, 231–32, 241–42
    animal model of, 200–204
    childhood and, 91–92
    deadliness of, 6
    definition of, 78
    gene expression and, 89–90, 200–205
    health and, 89–90, 231–32, 241–42,
        248

health care system and, 248
health risk factors and, 80–82
inflammation and, 89–90
intervention programs for, 230–31,
    233–39, 242–45
"loneliness epidemic," 147
"loneliness loop," 82
mortality and, 80, 232, 250
perception and, 78, 81–82, 242–43
as public health issue, 246–47
research on, 76–82
longevity, 6, 16–17, 135, 240, 246, 247
Lorenz, Konrad, 20, 21–22, 28, 49
Los Angeles's Unified School District,
    231
love (emotion), 39
low self-esteem, 101
loyalty, 93
lymph nodes, 202, 203

macaques. See rhesus macaques
Machiavellian intelligence hypothesis,
    119–20
MacLean, Paul, 45–46
Madingley Field Station, 29
magnetic resonance imaging (MRI), 75,
    82. See also functional magnetic reso-
    nance imaging (fMRI)
making friends, 91–115
mammals, 44, 97, 127. See also specific
    mammals
marital status, 232–33, 241–42, 247–48
    health care system and, 248
    mortality and, 233
    singledom, 241–42, 247–48
    widows/widowers, 232, 236, 246
Mark, 141, 149, 157, 184, 187
Marmot, Michael, 83–84
marriage, 148–49, 160. See also spouses
Mary, 98

Mass MOCA, 164
Massachusetts Institute of Technology, 53,
    62, 126
Matthew, 63–64, 65, 91–92, 94, 114, 115,
    187, 223, 251–52
Matthews, Sarah, 143
Max Planck Institute for Evolutionary
    Anthropology, 112, 218
Max Planck Institute for Human Cognitive
    and Brain Sciences, 211
McGill University, 14, 68
McGlone, Francis, 58
mean girls, 157, 177
medial prefrontal cortex, 209
medial temporal lobes, 209
Medicare, 248
Mendel, Gregor, 33
mentalizing, 47, 62–63, 208–9
mentalizing network, maturation of,
    62–63
mice, 205, 213
Michael (Mike), 183, 187
Michigan State University, 170, 174, 243
microscopic approach, 17
Miller, Greg, 204–5
mind-body connection, 74–75
mobile phones. See cell phones
mobility, 143, 170–71, 238
modern synthesis, 34
Moira, 8, 223
Mongolia, 146
Monitoring the Future, 168
monkeys. See primates
monocytes, 204
Montaigne, Michel de, 23, 156
moralistic aggression, 41
morality, 37–38, 42, 119
Moremi Game Reserve, 130–32, 135, 194,
    200, 240
Morris, D'Ann, 234

mortality
  loneliness and, 80, 232, 250
  marital status and, 233
  social isolation and, 71, 74
mother-child bond, 27–28, 30–31, 43–44,
    49–52, 58–59, 174–75. *See also* parents
multibrain interactions, 222–28
mutual gaze, infants and, 59
myelin, 105

National Heart, Lung, and Blood Institute,
    68
National Institute of Mental Health, 62,
    104–5
National Institutes of Health, 36–37
National Longitudinal Study of Adoles-
    cent to Adult Health (known as Add
    Health), 187–90
natural selection, 16–17, 39, 40, 128–29, 133
  friendship and, 130, 134–35
  social bonds and, 135–36
Nature Communications, 221–22
Nazi Germany, 42
necessity, as mother of invention, 128–29
neighbors, 86, 172
neocortex, 97
Nepal, 144, 205
nerve fibers, 57–58
nervous system, 87, 207
networks, personal, 4
neural connections, 104
neurological conditions, 6
neurons, 47, 56–57, 213
neuroscience, 7, 17, 24, 39, 43–64, 74–76,
    103–7, 112–13, 206–28
neuroticism, 94, 142
neurotransmitters, 87, 202, 217–28
Newcomb, Theodore, 14–15
New York Review of Books, 42
New York Times, 42, 178

New York University, 120
Nicholas, 63–64
Nin, Anaïs, 78
noradrenaline, 96
Northwestern University, 158, 204
nucleus accumbens, 47

occipital lobe, 46
Odgers, Candice, 177
Ohio State University, 74, 79, 84
Olausson, Håkan, 57–58, 59
older adults. *See also* aging
  living alone, 241–42
  social media and, 169–70
Ong, Wei Song, 214, 216
openness to experience, 94, 142
opportunity, 83–84, 139
Orben, Amy, 167–68, 178–79
orbitofrontal cortex, 209
outer circles, 150–55, 171–72
Oxford University. *See* University of Oxford
oxytocin, 112–13, 217–19, 220

pain network, 211
pairs, 102, 103
Pakaluk, Michael, 23
Panksepp, Jaak, 95, 96
parents, 150
  parent-child relationships, 28, 30–31,
    39, 43–44, 50–52, 59, 93, 174–75
  as social buffers, 112–15
parietal lobe, 46
Parkinson, Carolyn, 207, 212, 220–26
Parton, Dolly, 157
Passenger's Dilemma, 143–46
pediatricians, 179
peer relationships, 98–99, 107–11
  "peer effect," 107–10
  "peer presence," 110
  peer pressure, 110

perception
  loneliness and, 78, 242–43
  social isolation and, 78
  social life and, 78
personality, 142
  dimensions of, 94, 142
  personality factors, 94
PET scans, 75
Pew Research Center, 148, 149, 165, 170,
    172, 178, 242
phenotype, 187
philia, 24, 136
Phillips, Barbara, 234, 236
phone calls, 141–42, 150, 174. *See also* cell
    phones
physiology
  emotion and, 74
  relationships and, 84–86
  social factors and, 75–76
Piaget, Jean, 92
Pinker, Susan, 245–46
Pirazzoli, Laura, 45–46, 48, 56–57, 59
Plato, *Lysis*, 23–24
Platt, Michael, 3, 190–92, 196–97, 212–17,
    216, 218, 227
play, 93, 94–98
  "play histories," 96
  social play, 97–98
policy makers, 247
positive feeling, 24
posttraumatic stress disorder, 205
prairie voles, 218–19, 246
predictability, 240
prefrontal cortex, 47, 106, 113
Price, Wynona, 235–36, 238
Primack, Brian, 173
primates, nonhuman, 32, 44, 63, 111–12,
    116–20, 127, 129–30, 133, 136. *See
    also* primatology; *specific species*
  aging and, 240–41

baboons, 15, 116–17, 120, 128, 130, 131,
    134, 136–37, 176, 194, 197, 200, 246
chimpanzees, 32, 63, 111–12, 118, 119,
    120, 136, 176, 218
Machiavellian intelligence hypothesis
    and, 119–20
rhesus macaques, 1–4, 6, 9–11, 14–16,
    18, 29–30, 34–39, 54, 128, 190–204,
    212–17, 218, 246
social bonds and, 119–20
primatology, 7, 15, 29–30, 32, 119–20,
    121, 125–33, 135, 190–95, 198–200,
    240–41
Princeton University, 126, 226
*Proceedings of the National Academy of Sci-
    ences*, 187–90
"proof of concept," 224
prosocial behaviors, 62
prosopagnoisa, 53, 93
proteins, 203
proximity, 9, 14–15, 18, 137, 142, 146, 161,
    166, 194–97, 246
Przybylski, Andrew, 167–68, 180
psychoanalysis, 25
psychology
  biology and, 24–25, 87
  birth of, 24–25
psychopathy, 96
puberty. *See* adolescence
public health, 16, 67–68, 246–47. *See also*
    epidemiology
Punta Santiago, Cayo Santiago, 1–4

Ramón y Cajal, Santiago, 214
rank. *See also* hierarchy, vs. social bonds,
    134
rats, 213
Rauch, Jonathan, 239
Rawlins, Richard, 12
reciprocal altruism, 40–41

reciprocity, 15, 24, 40–41, 93
reconciliation, 119
rejection. *See* social rejection
relationships. *See also specific kinds of
relationships*
across the life span, 141–42
aging and, 87–88
ambivalence and, 85–86
complexity of, 146
definition of, 29
diversity of, 82–83
evolutionary success and, 130
health and, 65–90
measurement of, 5, 146
negative, 84–86, 250
persistence of, 170–72
physiology and, 84–86
positive, 84–86, 136, 250–51
protecting, 176
quality of, 135
quality vs. quantity of, 6, 84–85, 146
revival of dormant through social media,
170
social media and, 169
strength of, 135
relatives. *See* family members
reproductive success, 135, 197. *See also*
natural selection
reptiles, 44
reputation, 41
reward, 47
reward messages, 47
reward systems, 109–11, 214, 218
rhesus macaques, 1–4, 6, 9–11, 14–16, 18,
29–30, 34–39, 54, 128, 190–204,
212–17, 218, 246
Richard, 156
Ricks, Linda, 234–35, 236–37, 238, 239
Risa, 133
RNA, 88, 205. *See also* genetics
Robertson, James, 27, 29

romantic partners, 4. *See also* spouses
Royal Medico-Psychological Association
(today's Royal College of Psychiatrists), 20–22
Ruby, 133

sadness (emotion), 51
Sapolsky, Robert, 69, 105–6
Sara, 138, 139, 181
Sarämaki, Jari, 150
Sardinia, 245–46
Sarty, Merrill, 52
satisfaction, 204
Saxe, Rebecca, 62–63
Science, 72–73, 130, 134, 211, 218–19
Scientific Reports, 197
Sedgwick, Kyra, 151
Seeman, Teresa, 87–88, 231–33, 237,
238–39, 243, 246
self, sense of, 61–62, 208
self-control, 97
self-disclosure, 8–9, 154–55, 157, 161. *See
also* confidants
self–other overlap, 210–12
self-referential behaviors, 62
Selye, Hans, 68–69
senses, 45, 56–57. *See also specific senses*
sensory receptors, 57–58
serotonin, 217
serotonin pathways, 196
sexual mates, 136
Seyfarth, Robert, 121, 127–32, 134, 136,
190, 213, 218, 240
"shared intentionality," 63
sharing, 93
sheep, 7
Shensa, Ariel, 168–69, 179
Shlain, Tiffany, 175–76
shyness, 93–94, 183–205
siblings, 93, 149, 186–87
"sick while single," 247–48

Sierra, 131

Silk, Joan, 131–34, 194

similarity, 15, 222–23

Singer, Tania, 211, 219

singledom, 241–42, 247–48

SIV (simian immunodeficiency), 201–2

Six Degrees of Kevin Bacon, 151

sleep, decreased quality of, 80

Sloan, Robin, 133

Small, Mario Luis, 154–56

smartphones, 165, 169. *See also* cell phones

Smith, Adam, 24, 61

Smuts, Barbara, 130

Snapchat, 165, 177

sociability, 8–9

  in children, 59–60

  genes and, 183–205

social acceptance, 111

social behavior, 6–7, 8

  biology and, 190–92

  genes and, 207, 216–17

  heritability of, 183–205

  neuroscience and, 206–28

social bonds, 16–19, 247

  animals and, 87, 117–18

  natural selection and, 135–36

  primates and, 119–20

  quantifying dimensions of, 135

  vs. rank, 134

social brain

  adolescence and, 110–11

  brain lesions and, 216–17

  building a, 43–64

  circuitry of, 208–9

  development of, 207

  social brain hypothesis, 119–21

social buffering, 111–12, 113–15

social choices, 152–53

social cognition, evolutionary success and, 130

social communication, struggles with, 59–60

social connections, 71, 151, 166, 241, 249

  biological need for, 16

  building, 243–45

  continuity of, 171

  health and, 76–78, 238, 250–51

  interference caused by cell phones and, 174–75

  longevity and, 247

  nurturing of, 25

  as public health issue, 246–47

  through social media, 166, 169–70

social convoy, 146, 150, 248

social disconnection, 25

social dysfunction, 6

social evaluation, 114

social factors, physiology and, 75–76

social groups, 25

social hierarchy. *See* hierarchy; rank

social information, 51

social integration, 69–71, 76, 243–45

  health effects of, 80

  physiological effects of, 67

social isolation, 172, 232. *See also* loneliness

  dangers of, 82

  as a financial issue, 248

  health and, 70–71, 74, 80

  mortality and, 71, 74

  perception and, 78, 81–82

sociality, 121, 190–92, 208

sociality index, 134

socializing, 15–16, 183–84

social life

  measurement of, 71–72

  perception and, 78

social media, 23, 163–82, 178

  effects of, 166, 168–69

  intimacy and, 172–73

  positive effects of, 172

  research on effects of, 166–70

  teenagers and, 176–78

social mind, evolution of, 128–29

social network analysis, 146, 151–53, 188, 216–17
  animal research and, 191–95
  brain imaging and, 222–24
  emotion and, 153
social networking, 163–82
social networks, 171–74, 248. *See also* social media; social network analysis
  aging and, 239–41
  health and, 80
  heritability and, 188–89
  narrowing with age, 240
  technology and, 172–73
social neuroscience, 75–76
social play, 97–98. *See also* play
social primatology, 190–92. *See also* primatology
social psychology, 121–24
social rejection, 111
social relationships. *See* relationships
social sciences, 41
social skills, 118–19
  development of, 120–21
social smile, infants and, 59
social status, health and, 83–84
social stigma, force of, 89
social stimuli, adult brains and, 207–8
social support, 68–70, 146, 154, 177
  health care system and, 247–48
  importance of, 69, 74
  perception of, 86
  research on, 72, 73–74
  social media and, 170
social threat, perceptions of, 101
social ties, types of, 232–33
social withdrawal, 80
social world, diversity of, 82–83
sociobiology, 39–40, 41–42
Sociobiology Study Group, 42
sociology, 71, 147–48, 153–54. *See also* social network analysis

Socrates, 23–24, 166
somatosensory cortex, 58
soul mates, 149
Spitz, René, 26–27
spouses, 85, 86, 136, 148–49, 160
stability, 135, 142
Stanford Center on Longevity, 240
Stanford University, 61, 87, 167, 219, 240
  School of Medicine, 56
State University of New York at Stony Brook, 158–59
statistical methods, 168–69
status, 101–2. *See also* hierarchy
Steinberg, Ben, 107
Steinberg, Laurence, 107–11
Stephanie, 65, 70, 154, 210, 247
still-face paradigm, 175
strength (social network analysis), 188
"strength of weak ties," 153–54
stress, 68–69, 80, 85, 87, 112–14. *See also* fight-or-flight response
  adolescence and, 113–14
  aging and, 87–88
  caregivers and, 76
  evolutionary stress responses, 124–25
  force of, 89
  friendship and, 111–12
  gender and, 124–25
  health and, 201
  HIV virus and, 202–3
  immunology and, 75–76
  lowered by presence of friend, 7
striatum, 209
strong ties, 146, 154, 155–56
substantia nigra, 96
suicide, 25
superior temporal sulcus (STS), 47, 54–55, 209, 214, 216
support, 93, 157. *See also* social support
surprise (emotion), 51
Suzanne, 181

Sylvia (baboon), 131–32, 134, 176
symbiotic relationships, 40–41
Syme, Leonard, 67–68, 71, 72, 232
sympathetic nervous system, 204
sympathy, 41
synthetic families, 148

tactile cues, 49
Taylor, Shelley, 124–25
technology, 163–182
    children and, 179–80
    disruption caused by, 174–75
    effects of, 165, 166, 170–76
    mobility and, 170–71
    social networks and, 172–73
    teenagers and, 176–79
Technology Shabbat, 175–76
Tecumseh, Michigan, data from, 67–68, 72
teenagers. See also adolescence
    brain development of, 103–11, 207
    digital technology and, 174
    social media and, 165, 176–78
    technology and, 167–68, 176–79
    texting and, 178–80
Telfer, Elizabeth, 23
telomeres, 84, 86
temperament, 93–94
    average, 200
    high social, 200, 204
    low social, 200, 203–4
Temple University, 107
Templeton, Emma, 225
temporal cortex, 56
temporal lobe, 46
temporoparietal junction (TPJ), 47, 209
tend and befriend, 125
Teresa, Zia, 245–46
termites, 41
texting
    girls and, 178–79
    teenagers and, 178–80

theory of mind, 61–63, 207–8
"thick" friends, 146
three degrees of influence, 153, 224
"three-female experiment," 129–30
Tillich, Paul, 78
time, 24, 159–60
    devoted to friendships, 160–62, 248–49
    expenditure of, 240
    needed to make a friend, 160–62
    time constraints, 6
Time magazine, 42
Tinbergen, Nikolaas, 20, 28
toddlers, 60–61
togetherness, 7
Tom, 157
Tomasello, Michael, 63
Tomlin, Lily, 157
total peripheral resistance (TPR), 79
touch, 17, 56–57, 113
    affective, 58–59
    infants and, 56–59
TPH2, 196
transcription, 88–89, 201
transitivity (social network analysis), 188, 189
Travers, Michael, 234–35
"tribe years," 141
Trier Social Stress Test, 112
Trivers, Robert, 40, 41, 42
Trobriand Islanders, 144
trust (emotion), 41
trustworthiness, 93
"turnings" of life, 141
twin studies, 186–89

Uchino, Bert, 84–86, 87, 146–49
UCLA Loneliness Scale, 79–80, 200–201
Umberson, Debra, 72–73
uncertainty, 145–46
unfairness, 41
University of Alberta, 126

University of California, Berkeley, 67, 70, 96, 176
University of California, Davis, 198
University of California, Irvine, 177
University of California, Los Angeles, 31–32, 88, 99, 113, 126, 207
    Division of Geriatrics at the David Geffen School of Medicine, 231
University of Cambridge, 21, 30, 44, 49, 118, 127
University of Chicago, 79–80, 126, 200
University of Exeter, 2
University of Idaho, 97
University of Kansas, 157
University of Massachusetts Amherst, 198
University of Michigan, 14, 146, 148, 168
    Institute for Social Research, 71–72
University of Minnesota, 114
University of North Carolina, 243
University of Oxford, 28, 119, 167
University of Pennsylvania, 3, 130, 190, 212–17
    Medical Center, 212–17, 218
University of Pittsburgh, 168, 173
University of Puerto Rico, 3, 36–37
    Primate Research Center, 3–4, 9–11, 14, 15–16, 190–97, 216–17
University of Southern California, 52
University of Texas, Austin, 73
University of Utah, 84
University of Virginia, 122, 211
University of Winnipeg, 159
University of Wisconsin, 30, 112
Upswing Club, 239
USS Coamo, 35

Vaillant, George, 250
validation, 177
variation, 128–29
Venditti, Chris, 121

ventral striatum, 209
ventral tegmental area, 47, 96
ventromedial prefrontal cortex, 209
Vera, 154
vertebrates, 68
vervet monkeys, 129–30
video games, 17–18, 177–78
vision, 17, 45
visual cortex, 222–23
visual cues, 49, 177
voice, 55–56, 113
voles, 218–19, 246
voluntary groups, 172
von Frisch, Karl, 28
vulnerability, 8
Vygotsky, Lev, 92

Wake Forest University, 122
Waldinger, Robert, 250–51
Wall Street Journal, 248
Washington Post, 248
Watson, John B., 25–26
Watters, Ethan, 141
weak ties, 153–56
Weill Cornell Medical Center, 106
WEIRD (Western, educated, industrialized, rich, and democratic) societies, 143–44
well-being, 166–70, 173, 249
Wellington, Bertha, 236–37
Wellman, Barry, 166
Western Apaches, 144
whales, 136
Wharton, Edith, 209
Wheatley, Thalia, 207, 208, 212, 220–29
white blood cells, 204
Whitehall Studies, 83–84
white matter, 105
Whiten, Andrew, 119–20
widows/widowers, 232, 236, 246

Wiesel, Torsten, 103–4

William, 156

Williams, George, 40

Wilson, Edward O., 31–33, 34, 37–39, 41,
    42, 95

Wittig, Roman, 218

Wolf, Katherine M., 26

World Health Organization, 27

Wright, Sewall, 33–34

Wu, Paul, 52

Yale University, 113, 152, 226

yawning, 62

Young, Larry, 217–18

Zahn-Waxler, Carolyn, 62

Zak, Paul, 217

Zaki, Jamil, 61, 219

zebra fish, 6, 111–12, 246

zebras, 6, 136–37

Zimmer, Carl, 33–34